T0230719

HIGH-TECHNOLOGY DEVELOPMENT IN REGIONAL ECONOMIC GROWTH

High-Technology Development in Regional Economic Growth

Policy implications of dynamic externalities

BYUNG-ROK CHOI
Office of the Prime Minister, Government of Korea

Routledge
Taylor & Francis Group

LONDON AND NEW YORK

First published 2003 by Ashgate Publishing

Reissued 2019 by Routledge
2 Park Square, Milton Park, Abingdon, Oxon, OX14 4RN
52 Vanderbilt Avenue, New York, NY 10017

Routledge is an imprint of the Taylor & Francis Group, an informa business

Publisher's Note
The publisher has gone to great lengths to ensure the quality of this reprint but points out that some
imperfections in the original copies may be apparent.

Disclaimer
The publisher has made every effort to trace copyright holders and welcomes correspondence from
those they have been unable to contact.

A Library of Congress record exists under LC control number:

Typeset by Martingraphix, Cape Town, South Africa.

ISBN 13: 978-1-138-72517-1 (hbk)
ISBN 13: 978-1-315-19201-7 (ebk)

Contents

List of Tables

List of Figures

Acknowledgements

I would like to express my sincere appreciation to Dr. Brian J.L. Berry. This study could not have been completed without his guidance. Appreciation also goes to Drs. Ronald Briggs, James C. Murdoch, Paul Waddell, for their valuable comments and suggestions.

Special appreciation is expressed to Dr. Wim P.M. Vijverberg for his careful comments and suggestions on the econometric models; and to Dr. Lloyd J. Dumas for his helpful comments on the data arrangement; and to Dr. Heja Kim for her encouragement and suggestions. I am also indebted to my colleagues, Mrs. Nancy Juhn, Mr. Yeong-Seok Kim, and Mr. Hyo Park for their encouragement and comments. In addition, I would like to thank the Texas Workforce Commission for providing me with the recent data pertaining to Texas civilian labor force estimates.

Finally, I dedicate this study to my parents, Bong-Lim Choi and Yong-Sook Kim, and my mother in-law, Gap-Soon Lim. I deeply appreciate their encouragement and support during my years in the USA. Also my wife, Jung-Ran Lim, and my two daughters, Jee-Young (Jenny) and Jee-Eun (Jane); I am deeply appreciative of their sacrifice and prayers.

Chapter 1

Introduction

High-technology development is a process of business location, creation, and expansion that, since the 1980s, has contributed to the structural transformation and growth of many regional and local economies. As a result, many state and local governments have actively pursued policies and programs to encourage high-tech economic development. Much of the interest in high-tech industry has centered on its potential for high growth performance.

Because new forms of high-tech regional development have emerged, the phenomenon has received increasing attention in regional economics and economic geography. Much of the literature deals with questions of the location and growth of the industry and its impacts on regional development. As I demonstrate below, however, fundamental questions still remain unanswered.

High-tech firms are industrial 'organizations engaged in technically sophisticated activities that lead to product or process innovations, new inventions, or the creation of knowledge' (Goldstein et al., 1993: 148). Most current research uses aggregate indicators at the industry level to identify these high-tech firms. High-tech activity appears in a variety of production sectors, ranging from state-of-the-art products through routine products such as chemicals. The usual indicators include R&D intensity, technical workers (scientists, engineers, and often, technicians) as a percentage of the workforce, the rate of innovation and invention, or some combination.

Use of such indicators leads to innumerable difficulties, however (Malecki, 1991: 173–180). Different indicators result in different lists of high-tech industrial sectors. Aggregate indicators do not clearly reflect the innovative activities conducted by individual establishments. An alternative approach would be 'a more intuitive, but arbitrary, identification of science-based products and processes based on non-routine, state-of-the-art knowledge' at the establishment level (Malecki, 1991: 176). The point is acknowledged, but data constraints often prevent suitable analysis.

The geography of high-tech industry has changed significantly over time. According to Markusen et al. (1986: 129–31), the distribution of high-tech manufacturing employment is increasingly distinct from that of traditional industry and is highly uneven over regions, states and metropolitan areas. The newer sectors generally have different centers of agglomeration from the older ones in manufacturing. In maturing sectors, routinized activities have dispersed toward urban peripheries, toward smaller cities within the same regions, and interregionally. Scott et al. (1987: 216–17) illustrates these tendencies to agglomeration and dispersal in the electronics industry, perhaps the most representative high-tech sector. The question that arises is what forces are producing these new industrial patterns.

1

In the research reported in this study, I test three hypotheses about the nature of this high-tech industrial change, using the data for the state of Texas for the years 1988–1995:

1. Regional growth in high-technology employment is determined by endogenous growth processes (endogenous technological progress that embodies localization and urbanization, or specialization and diversity).
2. The growth of small, single-location firms is more likely to be driven by endogenous processes than is the growth of large, multi-locational firms.
3. Endogenous growth is more likely to occur in competitive local economies rather than in local economies dominated by one or relatively few large firms.

By testing these hypotheses, I believe that I am able to further understanding of the potential of high-technology development as an economic development tool and to assess which of two general approaches – attracting existing high-technology firms or encouraging regional business initiative – has the greater saliency.

The study is organized as follows. Chapter two provides the theoretical background. Chapter three outlines specific research questions and modeling strategy. Chapter four lays out empirical findings, and Chapter five discusses the policy implications.

Chapter 2

Theoretical Background and Policy Issues

Studies of high-tech development may be categorized into three major groupings: theoretical approaches regarding the sources of high-tech development, spatial and geographical analysis to identify locational preferences, and economic analysis to evaluate the economic effects of high-tech development. In the following sections, I first review these theories and related empirical findings. I then discuss the relevant policy issues.

2.1 Theories of High-technology Development

Many theories have been advanced to explain high-tech development. The principal distinctions are between long-wave theory and the associated product life-cycle paradigm, neoclassical growth theory and industrial location theory, cumulative causation theory and dynamic endogenous growth models, and theories of propulsive, innovative, and creative regions.

2.1.1 Long-wave Theory and the Product Life-cycle Paradigm

Long-wave theory suggests a recurring pattern of development over time. The key idea is expressed in Schumpeter's famous work, *Capitalism, Socialism and Democracy*. For Schumpeter (1950), capitalism is not stationary by nature. Technological revolutions periodically reshape the existing structure of industry, as capitalism develops along the long-wave, that is, the 50–55 year Kondratiev cycle.

According to Schumpeter, capitalism experiences periodic episodes of creative destruction. Old technology moves through a 50 year process of growth and saturation much like any other product life-cycle. Mature industries in the saturated market then are replaced by clusters of new technology. In a decade of rapid change, the economy goes into depression during which creative destruction occurs.

As suggested by this technological long-wave view, high-tech development is not necessarily a recent phenomenon. Different techno-economic systems have emerged at different moments in the history of capitalist development. For each regime, its specific high-tech industry, such as steel, automobiles, electricity, or computers, has been spawned by the clusters of innovations in each Kondratiev wave. Today, these older technological complexes are gradually declining, whereas new hi-tech venture-capital supported industry such as information technology is rapidly emerging (Pollard et al., 1996). This new industry is more footloose than older technological complexes.

3

Another important point of the long-wave view is that technological transition requires 'not merely clusters of hardware innovations but transformations of the entire socio-economic framework' (Hall et al., 1988: 25–26). In a locational context, it suggests that 'any industrial ensemble is likely in general to have somewhat different locational needs from previous ones' (Scott et al., 1987: 226). According to Berry (1991), the underlying, long-term, economic fluctuations are explicitly linked to political responses, although not determinate. Whether or not political institutions are favorable to business conditions depends on the accountability of the polity. For Berry, the key factors in economic development are the economic incentives for creative technological progress and the accountability of the political system.

The product life-cycle paradigm has been widely used to explain the locational patterns of manufacturing industries. More specifically in the context of high-tech industry, Markusen (1985) suggests that the profit cycle model be used to clarify the mechanisms that link profit level, industrial market structure, and the process of spatial division of labor. In Markusen's view, industrial sectors can be classified according to their stage in the profit cycle.

The product life-cycle model, and especially its regional variants, has undergone a great deal of criticism because of its oversimplifications (Malecki, 1991: 134–136). Despite many variations among products or industrial sectors, the basic notion of 'innovation', 'growth', 'maturity' and 'decline' seems to be useful for understanding the relationship between production stages, production organizations, industrial market structure and locations, however. The distinctions help not only to identify the determinants of high-tech development; they also provide meaningful policy suggestions. In particular, the product life-cycle paradigm is the most widely accepted explanation of the dispersal phenomenon (Barkley, 1988: 14).

2.1.2 Neoclassical Growth Theory and Industrial Location Theory

Neoclassical growth theory focuses on the equilibrium process in competitive markets. Output level is postulated as the aggregate production function of flexible inputs such as labor and capital. In the neoclassical system, as Solow (1956) explains, the economy has a steady-state equilibrium growth path, assuming constant returns to scale. The differences in regional growth rates primarily result from differences in production technologies. Over time, regional incomes converge as production factors freely flow among regions and the value of their marginal products equalizes.

Neoclassical models certainly have dominated economic growth theory since 1950s. Basically, however, output growth is conceived as a form of adjustment to factor supplies. Technology is assumed to be given. As a result, neoclassical growth theory has failed to explain a substantial part of the growth rate. Later studies show that the residual after quantifying contributions of factor inputs, referred to as technological progress, is in fact the most important part of output growth. Thus, many schools of thought have revolved around the objections regarding the assumptions and predictions of neoclassical growth theory.

The neoclassical approach to industrial location was pioneered by Weber (1929). Weberian location theory shows how the optimum location for an individual firm can be found in a simple situation, given distances to raw materials and the final market. The firm is assumed to minimize its total cost. Price levels are assumed to be exogenous. In Weber's simplified world, a specific industrial location is determined by the relative weights of three factors – transportation, labor costs, and agglomerative or deglomerative forces. In recent years, the Weberian framework has been expanded and generalized to encompass additional location factors and more complex circumstances.

For high-tech industry, there have been attempts to identify the most important location factors since the early 1980s (Joint Economic Committee, 1982; Office of Technology Assessment, 1984; Markusen et al., 1986; Harris, 1986; Hall, 1987; Ó hUallacháin et al., 1992). Most of these studies have concerned themselves with Weber's locational triangle, quality of life, agglomeration economies, policy measures, and so on. In contrast to classical location theory, studies of the location factors important to high-tech industry reveal that traditional locational factors such as materials and market pulls are relatively unimportant. Important factors commonly adduced in the literature include availability of highly-skilled labor (e.g., scientists and engineers), access to national and international markets (e.g., major airports), high-amenities (e.g., mild climates, educational and cultural advantages), agglomeration economies (e.g., concentration of high-tech firms, access to venture capital, specialized business services and a high degree of urban infrastructure), and socio-political variables (e.g., percent black, defense spending). Cost factors such as labor costs may be significant for mature or branch firms (Harris, 1986: 182).

Some factors remain in dispute. First, proximity to academic institutions may not be a necessary condition for high-tech development. Several studies have indicated the importance to high-tech industry of nearby research universities, especially those with major programs in science and engineering. Recent studies insist that proximity to academic institutions and science parks is one of high amenities (Shachar et al., 1992: 843). Surprisingly, the intensity of R&D (as measured by the volume of federal research contracts) is not significant (Hall, 1987: 153–154). Such concentrations would be significant only at certain period in time.

Secondly, quality of life variables tend to be tarnished in the long-run by migration and disagglomerative effects such as congestion and pollution. Using sectoral data for metropolitan areas for the period 1977–1984, Ó hUallacháin et al. (1992: 53) find that climate and recreation opportunities are seldom significant.

Finally, agglomeration economies, for example such externalities as shared information and shared skilled labor pool, are highly significant among high-tech firms (Markusen et al., 1986; Harris, 1986; Ó hUallacháin et al. 1992; Lyons, 1994). These external economies involve industry-specific (localization) or site-specific (urbanization) economies. While localization economies can arise from concentration of local firms in the same industry in a region, urbanization economies can arise from overall urban scale and diversity, that is, many different kinds of industry in large regions. However, what kinds of agglomeration economies are more important and to what extent or under what conditions is still a matter of dispute.

According to Calem et al. (1991: 83), when controlling for technical change (as measured by labor productivity), agglomeration economies (as measured by economies of scale) occur in urban areas with less than 2 million in population. Henderson (1986: 65–66) argues that for manufacturing industries, the relevant external economies of scale are those of localization, not urbanization. In contrast to

Henderson, Ó hUallacháin et al. (1992: 25) insist that both localization economies and urbanization economies are among the strongest determinants. For most of the manufacturing sectors and some services such as business services and engineering services, only localization economies are significant, but for many services, both localization and urbanization economies are significant. Harris (1986: 178) also finds that agglomeration economies are more important for new independent firms than for branch formations.

2.1.3 Cumulative Causation Theory and Dynamic Endogenous Growth Models

In neoclassical economic growth theory, technology is exogenous to local economic growth. Likewise, traditional locational theory is based on exogenous factors, assuming that firms respond to changes in the current comparative advantages of different locations. The central weakness of this static approach is that it cannot handle the evolutionary dynamics of innovative capitalist industry. Scott et al. (1987: 220) conclude that 'these exogenous locational factors turn out to be little more than ad hoc lists'. They suggest that meaningful theoretical generalizations to account for the emergence and expansion of the high-tech industry will be found in cumulative causation theory and endogenous growth models, in which technology is thought to be endogenous.

This thinking first appeared in Marshall (1890) and Young (1928), in Myrdal's regional growth concept (1957), in Verdoorn's law (1949), in Arrow's model (1962), and in Kaldor's model (1966). More recently, based on macroeconomic regularities, Kaldor (1978) argues that technological progress is endogenous to economic growth. To Kaldor, the cumulative, self-sustained nature of growth derives from the economies of scale in manufacturing, as expressed in the relationship known as Verdoorn's law (Choi, 1983: 155).

The modern version was formulated by Romer (1986) and Lucas (1988). According to Romer (1986), dynamic information externalities are the driving force in technological innovation and hence economic growth. In other words, technological progress associated with knowledge spillovers plays the most important role in regional economic growth. With technological progress, demand can be internally and repeatedly created in the economy. Thus, the economy grows cumulatively. Follain (1994) argues that technological externalities are an important example of these agglomeration economies. Such dynamic externalities are more readily observable in cities. Recent empirical analyses show that human capital investment in cities is particularly productive because of knowledge spillovers (Glaeser et al., 1994: 15–16).

With respect to knowledge spillovers, dynamic externalities have been hypothesized to take one of two possible forms, MAR (Marshall-Arrow-Romer) externalities and Jacobs externalities, based on the terminology used in Glaeser et al. (1991: 6–9). Both types involve information accumulation available to firms in a region. In MAR models, dynamic externalities arise from a buildup of knowledge associated with industry specialization. Agglomeration or concentration of the same industry in a region facilitates knowledge spillovers between firms through spying, imitation, and inter-firm labor movements. Historically, this kind of thought builds on Smith's (1776) division of labor, Marshall's (1890) industrial districts, Arrow's

(1962) learning by doing, and recently Romer's work (1986, 1987, 1990). On the other hand, in Jacobs' (1968) view, dynamic externalities arise from communications among firms in different industries, rather than specialization. Jacobs externalities are associated with diversity of local employment. According to Jacobs, the biggest innovations come from the diversified cities.

These two types of dynamic externalities are conceptually the counterparts of the two categories of static externalities, localization and urbanization, and have similar locational implications. But, whereas static externalities have been used to explain industrial location patterns, dynamic externalities have broader, evolutionary implications (Glaeser et al., 1991; Henderson et al., 1995). With dynamic externalities, we can explain industrial development over time.

Another important dimension of dynamic externalities concerns the conditions under which information externalities are maximized. The related discussions include social factors (e.g., mixture of social classes, education level and social stability) and political factors (e.g., optimal size and functional distribution of cities, and coordination) as well as economic factors. On the economic side, in particular, there is the debate between local monopoly and competition. The issue is the impact of local market structure on technological progress and economic growth.

In MAR theory, local monopoly is better for growth than local competition. According to Romer (1990), local monopoly allows innovators to internalize externalities and only internalization can propel innovation and growth. Otherwise, the introduction of the innovation may not be worthwhile. On the other hand, Porter (1990) insists that MAR externalities can be maximized when local industries are competitive. For Porter's model, local competition and rapid churning generate more innovation and imitation than local monopoly. Jacobs also argues for local competition. In Jacobs' view, monopoly tends to suppress economic growth, but local competition stimulates innovation (Jacobs, 1984: 227).

Whether regional economic growth is endogenous or not is an important question in terms of both theory and policy. If technological progress is purely an endogenous phenomenon, dynamic externalities can produce long-term divergence in economic growth. This is the opposite prediction to neoclassical theory. Under these conditions, effective economic development policy needs to be delivered locally. Moreover, since initial conditions matter, a contingent approach that considers the local context may be appropriate. Empirical evidence indicates that endogenous technological progress provides a much more convincing explanation for cumulative, self-propelling growth of the capitalist economy. Investigators now agree that the location and growth patterns of high-tech industry are influenced by the historical industrial environment in a region (Glaeser et al., 1991; Miracky, 1992; Henderson et al., 1995). However, many other questions are still unsettled and the empirical results conflict.

Glaeser, Kallal, Scheinkman and Shleifer (1991) examine knowledge spillovers within metropolitan areas. They report a cross-section regression analysis of employment growth in six large mature 2-digit industries across 170 standard metropolitan statistical areas (SMSAs) in 1956 and 1987. The results indicate that city-industries grow faster when the city is not highly specialized, and when the rest of the city is less specialized in the initial year. They interpret these findings to mean that important knowledge spillovers might be between rather than within industries.

These results are inconsistent with the theories of MAR externalities and Porter, but consistent with the theory of Jacobs externalities. Another interesting finding is that local competition (as measured by establishments per employee in the city-industry relative to those in the US industry in 1956) promotes growth. This result supports Jacobs' view.

Jaffe, Trajtenberg, and Henderson (1993) focus on the question of whether and to what extent knowledge spillovers are geographically localized. As evidence, they compare the geographic location of patent citations and that of the cited patents between 1975, 1980 and 1989. They find that citations are more likely to come from the same state and SMSA as the cited patents. These localization effects are quite large and significant at the local level and localization effects fade very slowly over time.

In contrast to Glaeser et al. (1991), Miracky (1992) finds evidence of both MAR externalities and Jacobs externalities, using data for manufacturing industries in 1977 and 1987. Henderson, Kuncoro, and Turner (1995) also confirm his finding. They use data for eight manufacturing 2-digit industries across 224 metropolitan areas in 1970 and 1987. In particular, to examine developmental implications regarding dynamic externalities, they compare five traditional capital goods industries (e.g., primary metals, machinery, electrical machinery, transport equipment, and instruments) which are wide-spread across cities, with three high-tech industries (e.g., computers, electric components, and medical equipment). To avoid selection bias, they use discrete-continuous choice analysis and joint maximum likelihood estimation method.

The results indicate that there are only MAR externalities for mature capital goods industries, and both MAR externalities and Jacobs externalities for new high-tech industries. They argue that new industries prosper in large metropolitan areas with a history of industrial diversity, but with maturity, production lines subject to MAR externalities eventually decentralize to smaller, more specialized cities, with lower wage and land costs. They claim that these results are consistent with product life-cycle notions.

Henderson (1994) also examines the role of MAR externalities versus Jacobs externalities, focusing on the lag structure of externality measures. Unlike the previous studies, he uses 11-year panel data for four 2-digit industries (e.g., chemicals, primary metals, machinery, and electrical machinery) across 742 urban counties for the period of 1977 to 1987. This study assumes that a location fixed/random effect could generate the role of history. The fixed/random effects model is estimated by the generalized method of moments procedure (e.g., a generalization of three-stage least squares or full information estimation).

The results indicate that both MAR and Jacobs externalities are present for manufacturing industries. Henderson argues that employment growth for manufacturing industries also is stimulated both by preexisting concentration and by preexisting diversity, and that diversity is important not only in attracting the industries to a region, but also in retaining industries. With respect to lag structure, he finds that historical concentrations of own industry affect employment levels for the next 5 to 6 years, whereas diversity effects persist beyond the 6-year time horizon. The lag structure of both measures shapes an inverted quadratic curve.

These empirical analyses provide conflicting evidence on the role of diversity and specialization and their time lags. The issue is the extent to which both technological externalities are important for high-tech industry. Despite variation by industry and by industrial production stage, depending on data and methodology, it can be argued that at least for high-tech industry, both externalities are likely to be crucial. Glickman (1994: 118) insists that this idea is consistent with many successful local economic development policies for high-tech industry.

But, product life-cycle notions of dynamic externalities for newer and mature industries are still unclear. More investigation on this issue might help clarify what kind of externalities have an advantage in attracting newer industries. It can be hypothesized that there is a relationship between dynamic externalities, production stage, and organizational type (e.g., single or multi-locational). For example, the growth of small, single-location firms is more likely to be driven by endogenous processes than is the growth of large, multi-locational firms.

In terms of conditions under which dynamic externalities could be effective, Glaeser et al. (1991) show that local competition encourages employment growth in industries. However, other relevant literature (Mansfield, 1994: 536–537) reveals that the answer is not so clear-cut. For high-tech industry, we lack empirical analysis. On the other hand, the interaction between knowledge spillovers and education level seems to be significant. The literature indicates that cities with higher concentrations of skilled workers and professionals have higher probabilities of having high-tech industry (Henderson et al., 1995: 1083).

2.1.4 Theories of Propulsive, Innovative, and Creative Regions

Theories of propulsive, innovative, and creative regions focus on structural reasons leading to long-term divergence of regional economic growth. These disequilibrium models are classified into two groups (Goldstein et al., 1993: 155–158). The first group (e.g., growth pole theory and innovation diffusion theories) stresses diffusion of growth or innovation, whereas the second group (e.g., seedbed/incubator theories) emphasizes nonspatial synergies with environmental factors.

It is difficult to analyze the specific role of technological change with these frameworks. Nevertheless, these theories correctly point out the importance of propulsive key industries and of creative individuals such as entrepreneurs, scientists and engineers. However, initial social and political imbalances can result in different patterns of growth. In fact, historical studies show that high-tech economic development in new industrial regions is not a random process. High-tech economic development seems likely to be strongly affected by social and political factors. Among other things, the location of military research facilities and defense spending has been important factor for high-tech development in the United States (Hall, 1987: 153).

2.2 Policy Issues: Impacts of High-tech Development on Regional Economic Growth

In the past few decades, local economic development has made considerable progress due to political decentralization and economic restructuring. According to one recent survey (Prager et al., 1995: 29–30), 41% of all cities and counties report the use of a formal economic development plan. Economic development planning is far more common among local governments serving large populations. Increasingly, urban economic development strategy tends to focus on high-tech development (Shachar et al., 1992: 839–840; Muniak, 1995: 803). The successful stories of high-tech regions such as Silicon Valley have encouraged regional governments to promote similar high-tech development in other regions. Also, high-tech development has been thought to be some form of relief from structural crises in the American economy during the 1980s. As a result, among state and local governments, there has been the intense competition for high-tech development.

Such efforts for high-tech development are ultimately designed to influence the direction of the local economy by fostering growth. Recent studies have drawn attention to the economic performance (e.g., economic growth, innovativeness, and the gross amount of induced investment) of high-tech industry, the relationship between high-tech activities and the regional economy (e.g., changes in industrial mix and occupational structure, and technological improvements and productivity gains in existing establishments), and more broadly, several social and political issues associated with high-tech development (e.g., bifurcation of occupational structure and associated political and social shifts, service and facility shortfalls, and groundwater pollution). Many of these lie beyond the scope of this study, focusing on the effects of high-tech industry on regional economic growth.

2.2.1 High-tech Industry as a Regional Economic Development Tool

The major policy issue addressed in this study is whether there exists a positive relationship between high-tech development and regional economic growth. At one level, high-tech industry is perceived as a source of new employment. However, since at the same time new technology also restructures other industries, dislocations must be taken into account to evaluate the high-tech industrial performance. On the whole, the evidence has tended to support the job creation potential of high-tech industry, although the growth rates vary among industrial sectors (Markusen, Hall, and Glasmeier, 1986). More jobs are created by technology than are eliminated (Cyert and Mowery, 1987). The negative impact of technology falls heavily on those without basic skills and with poor educations.

At the regional level, however, the literature has not been so clear about whether high-tech industry can be a dependable economic base (Malecki, 1991: 54–55). Some empirical evidence suggests that the effects of high-tech growth on regional employment are weak (Shachar et al., 1992: 840–843). Some high-tech clusters, such as the research triangle of North Carolina, have not acted as growth poles for regional growth.

2.2.2 The Role of Government in High-tech Economic Development

A major question is whether policy intervention can enhance high-tech economic development. If so, what public policies are likely to be effective? According to Malecki (1989: 74–75), state and local government should address the important economic foundations – that is, human resources, education, and quality of life – of their economies. The reason for this is that such components as industry mix, availability of venture capital, and entrepreneurship are hard to change in the short-run.

At first blush, this argument is convincing. Technological capability is the core of economic development. Once-established, technology tends to be inert, and fundamental change of technology is a long-term process. Low tax rates and local public expenditures are seldom significant (Office of Technology Assessment, 1984: 104; Harris, 1986: 182; Ó hUallacháin and Satterthwaite, 1992: 25). Economic development incentives such as research parks and enterprise zones have a marginally significant effect on local employment growth, although focused development incentives that emphasize infrastructure improvements may have some effects (Ó hUallacháin and Satterthwaite, 1992: 54).

These arguments are, in effect, against business attraction programs which market localities to attract desirable industry. However, recent surveys show that on average, business attraction programs remain more widely utilized than business retention programs in local economic development planning, although emphasis is increasingly on business retention programs (Prager et al., 1995: 31). This study will assess the validity of those two general approaches to promoting high-technology economic development – attracting existing high-technology firms and encouraging regional business initiative.

Chapter 3

Research Questions and Modeling Strategy

Several studies have examined the effects of dynamic externalities on regional economic growth. This study deals with theoretically arguable hypotheses regarding the nature of high-tech development. In this chapter, I first suggest three major hypotheses. I then review the empirical model appropriate to analyze the dynamics of high-tech industry and specify the regression models, such as cross-sectional and time-series models, to explore the differing effects on regional high-tech growth by firm size, organizational type, and product. I then discuss the research design strategy that is employed in the regression estimations. Finally, I review the empirical model to analyze the effects of high-tech development on regional economic growth.

3.1 Research Questions

Three major research questions are suggested by the literature just reviewed:

1. Whether regional growth in high-tech employment is determined by endogenous growth processes (endogenous technological progress that embodies localization and urbanization, or specialization and diversity).
2. Whether the growth of small, single-location firms is more likely to be driven by endogenous processes than is the growth of large, multi-locational firms.
3. Whether endogenous growth is more likely to occur in competitive local economies rather than in local economies dominated by one or relatively few large firms.

In what follows I treat these questions as hypotheses, and I test them using data for the state of Texas.

3.2 Conceptual Framework

The conceptual framework within which the hypotheses will be tested is outlined in Figure 3.1. I hypothesize that regional high-tech growth (employment or output) is determined by dynamic externalities, controlling for other regional characteristics. Just as in Kaldor's scheme, a feedback relationship is hypothesized between regional high-tech growth and dynamic externalities.

These dynamics can be captured in the following model:

For a product k, in location i at time t,

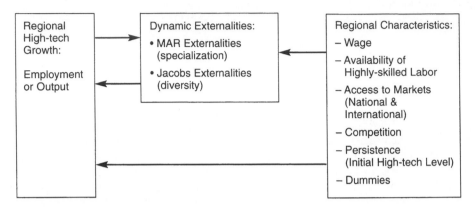

Figure 3.1 Conceptual Framework

$$YGk(it) = f(\sum_{m=0}^{6} SPk(i,t-m), \sum_{m=0}^{6} Dk(i,t-m), Wk(i,t), X(il), Yk(il), DM) \qquad (1)$$

$$SPk(it) = g(TYGk(it), \sum_{m=0}^{6} Wk(i,t-m), X(il), Yk(il), DM) \qquad (2)$$

$$Dk(it) = h(TYGk(it), \sum_{m=0}^{6} Wk(i,t-m), X(il), TYGo(it), Yk(il), Yo(il), DM) \qquad (3)$$

where $Yk(it)$, $Yk(i,t-1)$ = employment of product k at time t and time $t-1$, respectively,
$YGk(it) = Yk(it) - Yk(i,t-1)$ = employment growth of product k at time t,
$SPk(i,t-m)$ = specialization index of product k with time-lag $m=0...6$,
$Dk(i,t-m)$ = diversity index of product k with time-lag $m=0...6$,
$Wk(i,t-m)$ = wages with time-lag $m=0...6$,
$X(il)$ = availability of highly-skilled labor, access to markets, and competition,
$Yk(il)$, $Yo(il)$ = persistence (initial level) of k product or other high-tech products,
$TYGk(it) = Yk(it) - Yk(il)$ = total employment growth of product k at time t,
$TYGo(it) = Yo(it) - Yo(il)$ = total employment growth of other products at time t,
DM = dummies for regions (e.g., CMSA or MSA).

In equation (1), I postulate that for a high-tech product, regional employment growth is a function of dynamic externalities, that is, either MAR externalities (measured by specialization) or Jacobs externalities (measured by diversity), wage, and other regional characteristics. In this equation, specialization and diversity are assumed to have the distributed lags of $m=0...6$. These effects seem to persist for some period, forming dynamic endogenous technological processes. Also, wage is assumed to have time lags of $m=0...6$. Wage effects seem to persist for some period of time, because firms adjust scale and factor proportions to changes in wages. Higher wages are likely to reduce high-tech employment, other things being equal.

Other regional characteristics are controlled. Such control variables are measured by their initial level. These include availability of highly-skilled labor, access to national and international markets (e.g., airport hub access), competition (e.g., ratio of establishments per worker relative to those in the state of Texas), and persistence (initial high-tech level). Since there could be a simultaneous relationship between infrastructure development, job growth and the provision of other public goods, the beginning infrastructure endowment needs to be controlled (Ó hUalláchain and Satterthwaite, 1992: 39). However, instead of including an infrastructure variable, regional dummies will be used to capture all the other factors that are not explicitly included.

In equation (2) and equation (3), current dynamic externalities – that is, either MAR externalities (measured by specialization) or Jacobs externalities (measured by diversity) – are formulated to be a function of high-tech employment growth, wage, and other regional characteristics. Unlike specialization, diversity for product k is assumed to depend on the initial employment level and growth of other high-tech products, as well as on those of high-tech product k. Wage is also assumed to have time lags of $m=0...6$.

In this model, at time t, such variables as regional high-tech employment growth (YGk), total high-tech employment growth (TYGk), current specialization (SPk), and current diversity (Dk) are endogenous and simultaneously determined. As in Kaldor's scheme, high-tech development seems to be propelled by the dynamic relationship between employment (or output) growth and productivity change. In the model specification, regional high-tech growth and dynamic externalities are interrelated, involving economies of scale and technological progress. Past specialization and past diversity are assumed to be exogenous variables at time t. Other variables such as wages and other regional characteristics are also thought to be exogenous variables at least over the time period studied. Because they may be endogenous from a longer time perspective, I assume that these variables are not strictly exogenous, but merely predetermined.

In the model, the equations are estimated by the two-stage least squares estimation method. As a statistical criterion, the Hausman specification test is employed to choose between a simultaneous equation model that permits simultaneity and one that does not. In order to derive policy implications, the model also is estimated for the different types of organizational forms (e.g., small and large, single and multi-locational firms).

3.3 Operational Form

The purpose of this section is to operationalize the conceptual framework. I first specify the regression models by means of which major hypotheses are tested, and then define the variables to be used for estimation.

Hypothesis 1 is that:

> Regional growth in high-tech employment is determined by endogenous growth processes.

This hypothesis is tested first cross-sectionally. For total high-tech industry, the three equations can be rewritten operationally as:

$$TYG95 = f (SP88 \; D88 \; MW88$$
$$COLL \; ACCESS \; COMP88 \; Y88 \; CMSA \; MSA) \qquad (4)$$
$$SP95 = g (TYG95 \; MW88$$
$$COLL \; ACCESS \; COMP88 \; Y88 \; CMSA \; MSA) \qquad (5)$$
$$D95 = h (TYG95 \; MW88$$
$$TYGOM95 \; YOM \; COLL \; ACCESS \; COMP88 \; Y88 \; CMSA \; MSA) \qquad (6)$$

The Operational definitions of the variables are summarized in Table 3.1. A few things should be noted. For total high-tech employment growth (TYG95), since many locations in 1988 have zero activity, employment change of high-tech industry is used instead of a rate of change measure. The specialization index (SP88, SP95) represents relative concentration within the state of Texas. A value greater than 1 means that location i is more specialized in high-tech industry than the average in the state of Texas. The prediction of MAR theory is that high specialization of high-tech industry in location i should speed up growth of high-tech industry in that location. The measure of diversity (D88, D95) is constructed for all high-tech activities. An increase in this diversity index reflects less diversity. The maximum of this variable is 1, whereas the minimum is zero. According to Jacobs' theory, a lower value of this measure of diversity should be associated with faster high-tech growth. Local competition (COMP88) is measured by the number of establishments per worker in high-tech industry in location i relative to the number of establishments per worker in high-tech industry in the state of Texas. A value greater than 1 means that high-tech industry in location i is more competitive than the average for the state of Texas (Glaeser et al., 1991: 16).

The second step is to test hypothesis 1, using the time-series analysis and simultaneous equation model. To reduce the severity of multicollinearity, because the levels of specialization and diversity and the wage variables are highly correlated, a first difference transformation is employed. There is a loss of one observation due to the differencing procedure. In addition to high-tech employment growth (YG95), total high-tech employment growth (TYG95), current specialization (SP95), and current diversity (D95), the endogenous variables are the yearly change of specialization (DSP95) in 1995, and the yearly change of diversity (DD95) in 1995.

Table 3.1 Operational Definitions of the Variables:
Cross-sectional Analysis for High-tech Industry

Code	Variable	Operational Definition
TYG95	total high-tech employment growth	employment change of high-tech industry location i between 1988 and 1995
SP88, SP95	specialization index for high-tech industry	ratio of high-tech employment per manufacturing industrial employment in location i relative to those in the state of Texas for the years 1988 and 1995
D88, D95	diversity index (Hirschman-Herfindahl Index) for high-tech industry	Σ (employment share for j)5 Σ = (employment for j / total high-tech employment)5 for any j high-tech product in location i for the years 1988 and 1995
MW88	manufacturing wage	annual average payroll for manufacturing industries in 1988 in the county where location i locates
COLL	availability of highly-skilled labor	percentage of the population more more than 25 years old with at least college degree in 1990
ACCESS	access to national and international markets	road distance of location i to airport hub (either Dallas-Fort Worth or Houston)
COMP88	local competition for high-tech industry	ratio of establishments per worker in location i relative to those in the state of Texas in 1988
Y88	persistence for high-tech industry	initial high-tech employment level in location i in 1988
TYGOM95	total employment growth of other manufacturing industries	employment change of manufacturing industries between 1988 and 1995 in the county where location i locates
YOM88	persistence of manufacturing industries	initial level of manufacturing industries in 1988 in the county where location i locates
CMSA	dummy for Consolidated Metropolitan Statistical Area	1 if location i is located in CMSA 0 if not
MSA	dummy for Metropolitan Statistical Area	1 if location i is located in MSA 0 if not

The three equations become:

$$YG95 = f\,(DSP95\ DSP94\ DSP93\ DSP92\ DSP91\ DSP90\ DSP89$$
$$DD95\ DD94\ DD93\ DD92\ DD91\ DD90\ DD89$$
$$MW95\ COLL\ ACCESS\ COMP88\ Y88\ CMSA\ MSA) \tag{7}$$

$$SP95 = g\,(TYG95$$
$$DMW95\ DMW94\ DMW93\ DMW92\ DMW91\ DMW90\ DMW89$$
$$COLL\ ACCESS\ COMP88\ Y88\ CMSA\ MSA) \tag{8}$$

$$D95 = h\,(TYG95$$
$$DMW95\ DMW94\ DMW93\ DMW92\ DMW91\ DMW90\ DMW89$$
$$TYGOM95\ YOM88\ COLL\ ACCESS\ COMP88\ Y88\ CMSA\ MSA) \tag{9}$$

In equation (7), only current manufacturing wage (MW95) is included, assuming that there is an equilibrium between current wage and high-tech employment. In this sense, inclusion of the successive differences of manufacturing wage variables could be redundant.

In this simultaneous equation model, there are three identities:

$$TYG95 = Y95 - Y88$$
$$DSP95 = SP95 - SP94$$
$$DD95 = D95 - D94$$

The remaining variables are assumed to be predetermined in the model. Solution is by the two-stage least-squares estimation method, with a Hausman specification test. Table 3.2 summarizes the operational definitions of the new variables. A positive value of the yearly change of specialization variables (DSP) denotes an annual increase of specialization, whereas a positive value of the yearly change of diversity variables (DD) means an annual decrease of diversity.

Table 3.2 Operational Definitions of the Variables:
Time-series Analysis for High-tech Industry

Code	Variable	Operational Definition
YG95	high-tech employment growth	employment change of high-tech industry in location i between 1994 and 1995
DSP95 DSP94 DSP93 DSP92 DSP91 DSP90 DSP89	yearly change of specialization for high-tech industry	DSP95=SP95–SP94 DSP94=SP94–SP93 DSP93=SP93–SP92 DSP92=SP92–SP91 DSP91=SP91–SP90 DSP90=SP90–SP89 DSP89=SP89–SP88

DD95 DD94 DD93 DD92 DD91 DD90 DD89	yearly change of diversity (Hirschman- Herfindahl Index) for high-tech industry	DD95=D95–D94 DD94=D94–D93 DD93=D93–D92 DD92=D92–D91 DD91=D91–D90 DD90=D90–D89 DD89=D89–D88
DMW95 DMW94 DMW93 DMW92 DMW91 DMW90 DMW89	yearly change of manufacturing wage	DMW95=MW95–MW94 DMW94=MW94–MW93 DMW93=MW93–MW92 DMW92=MW92–MW91 DMW91=MW91–MW90 DMW90=MW90–MW89 DMW89=MW89–MW88

NOTE: The lagged variables of specialization and diversity are defined as in Table 3.1.

Hypothesis 2 is that:

> The growth of small, single-location firms is more likely to be driven by endogenous processes than is the growth of large, multi-locational firms.

An important role often asserted for small independent firms is that they function as the seedbed for new enterprises capable of challenging established businesses. The greater presence of independents in a location is also thought to be evidence of the seedbed effect often discussed in connection with the high-tech industry. For policy implications, it seems to be desirable to distinguish small vs. large firms and single vs. multi-locational establishments. Small firms are defined as establishments under 50 persons on the average from 1988 to 1995. Single firms are those establishments that are a single headquarters establishment. Multi-locational establishments include branches or subsidiaries owned by a headquarters outside the same location.[1]

To verify hypothesis 2, high-tech employment in location i needs to be disaggregated into separate groups (small vs. large firms and single vs. multi-locational establishments). Mathematically:

$$
\begin{aligned}
Y95 &= YSM95 &+ YL95 &= YSI95 &+ YM95 \\
YG95 &= YGSM95 &+ YGL95 &= YGSI95 &+ YGM95 \\
TYG95 &= TYGSM95 &+ TYGL95 &= TYGSI95 &+ TYGM95
\end{aligned}
$$

where Y95, YSM95, YL95, YSI95, YM95 = high-tech employment of all, small and large firms, single and multi-locational establishments, respectively, in location i in 1995; YG95, YGSM95, YGL95, YGSI95, YGM95 = high-tech employment growth of all, small and large firms, single and multi-locational establishments, respectively, in location i, between 1994 and 1995;

TYG95, TYGSM95, TYGL95, TYGSI95, TYGM95 = total high-tech employment growth of all, small and large firms, single and multi-locational establishments, respectively, in location i between 1988 and 1995.

Using the above notation, for each paired-group, cross-sectional and time-series models are developed. For small and large firms, the cross-sectional model to test hypothesis 2 is as follows:

$$
\begin{aligned}
\text{TYGSM95} \quad &= f\,(\text{SP88 D88 MW88} \\
&\qquad \text{COLL ACCESS COMP88 Y88 CMSA MSA}) \qquad (10) \\
\text{TYGL95} \quad &= g\,(\text{SP88 D88 MW88} \\
&\qquad \text{COLL ACCESS COMP88 Y88 CMSA MSA}) \qquad (11) \\
\text{SP95} \quad &= h\,(\text{TYGSM95 TYGL95 MW88} \\
&\qquad \text{COLL ACCESS COMP88 Y88 CMSA MSA}) \qquad (12) \\
\text{D95} \quad &= i\,(\text{TYGSM95 TYGL95 MW88} \\
&\qquad \text{TYGOM88 YOM88 COLL ACCESS COMP88 Y88 CMSA} \\
&\qquad \text{MSA}) \qquad\qquad\qquad\qquad\qquad\qquad\qquad (13)
\end{aligned}
$$

In equations (10) and (11), total high-tech employment growth of small and large firms (TYGSM95 and TYGL95), are formulated to be a function of initial level of specialization (SP88) and diversity (D88), controlling for other regional characteristics. In equations (12) and (13), current specialization and diversity (SP95 and D95) are formulated to be a function of high-tech employment growth of small and large firms (TYGSM95 and TYGL95), controlling for other regional variables and dummies. With these equations, the differences by firm size can be identified. All the explanatory variables and their operational definitions are the same as in the cross-sectional model for total high-tech industry.

The time-series model for total high-tech industry is also expanded to distinguish between small and large firms. The simultaneous equation model is composed of four equations as follows:

$$
\begin{aligned}
\text{YGSM95} \quad &= f\,(\text{DSP95 DSP94 DSP93 DSP92 DSP91 DSP90 DSP89} \\
&\qquad \text{DD95 DD94 DD93 DD92 DD91 DD90 DD89} \\
&\qquad \text{MW95 COLL ACCESS COMP88 Y88 CMSA MSA}) \qquad (14) \\
\text{YGL95} \quad &= g\,(\text{DSP95 DSP94 DSP93 DSP92 DSP91 DSP90 DSP89} \\
&\qquad \text{DD95 DD94 DD93 DD92 DD91 DD90 DD89} \\
&\qquad \text{MW95 COLL ACCESS COMP88 Y88 CMSA MSA}) \qquad (15) \\
\text{SP95} \quad &= h\,(\text{TYGSM95 TYGL95} \\
&\qquad \text{DMW95 DMW94 DMW93 DMW92 DMW91 DMW90 DMW89} \\
&\qquad \text{COLL ACCESS COMP88 Y88 CMSA MSA}) \qquad (16) \\
\text{D95} \quad &= i\,(\text{TYGSM95 TYGL95} \\
&\qquad \text{DMW95 DMW94 DMW93 DMW92 DMW91 DMW90 DMW89} \\
&\qquad \text{TYGOM95 YOM88 COLL ACCESS COMP88 Y88 CMSA MSA}) \\
&\qquad\qquad\qquad\qquad\qquad\qquad\qquad\qquad\qquad (17)
\end{aligned}
$$

Besides, there are four identities:

TYGSM95 = YSM95–YSM88
TYGL95   = YL95–YL88
DSP95   = SP95–SP94
DD95   = D95–D94

In this model, the endogenous variables are high-tech employment growth (YGSM95 and YGL95), total high-tech employment growth (TYGSM95 and TYGL95), current specialization and diversity (SP95 and D95), and the differences of specialization and diversity (DSP95 and DD95). As mentioned, high-tech employment growth and dynamic externalities are currently and simultaneously determined. The simultaneous equation model is identified by the two-stage least-squares estimation method and tested by a Hausman specification test.

By the same token, cross-sectional and simultaneous equation models for single and multi-locational establishments also are developed to test hypothesis 2. The cross-sectional model is summarized as follows:

$$
\begin{aligned}
\text{TYGSI95} \;&= f\,(\text{SP88 D88 MW88} \\
&\qquad \text{COLL ACCESS COMP88 Y88 CMSA MSA}) &(18) \\
\text{TYGM95} \;&= g\,(\text{SP88 D88 MW88} \\
&\qquad \text{COLL ACCESS COMP88 Y88 CMSA MSA}) &(19) \\
\text{SP95} \;&= h\,(\text{TYGSI95 TYGM95 MW88} \\
&\qquad \text{COLL ACCESS COMP88 Y88 CMSA MSA}) &(20) \\
\text{D95} \;&= i\,(\text{TYGSI95 TYGM95 MW88} \\
&\qquad \text{TYGOM95 YOM88 COLL ACCESS COMP88 Y88 CMSA MSA}) \\
& &(21)
\end{aligned}
$$

The time-series analysis and simultaneous equation model includes four equations and four identities as follows:

$$
\begin{aligned}
\text{YGSI95} \;&= f\,(\text{DSP95 DSP94 DSP93 DSP92 DSP91 DSP90 DSP89} \\
&\qquad \text{DD95 DD94 DD93 DD92 DD91 DD90 DD89} \\
&\qquad \text{MW95 COLL ACCESS COMP88 Y88 CMSA MSA}) &(22) \\
\text{YGM95} \;&= g\,(\text{DSP95 DSP94 DSP93 DSP92 DSP91 DSP90 DSP89} \\
&\qquad \text{DD95 DD94 DD93 DD92 DD91 DD90 DD89} \\
&\qquad \text{MW95 COLL ACCESS COMP88 Y88 CMSA MSA}) &(23) \\
\text{SP95} \;&= h\,(\text{TYGSI95 TYGM95} \\
&\qquad \text{DMW95 DMW94 DMW93 DMW92 DMW91 DMW90 DMW89} \\
&\qquad \text{COLL ACCESS COMP88 Y88 CMSA MSA}) &(24) \\
\text{D95} \;&= i\,(\text{TYGSI95 TYGM95} \\
&\qquad \text{DMW95 DMW94 DMW93 DMW92 DMW91 DMW90 DMW89} \\
&\qquad \text{TYGOM95 YOM88 COLL ACCESS COMP88 Y88 CMSA MSA}) \\
& &(25)
\end{aligned}
$$

TYGSI95 = YSI95–YSI88
TYGM95 = YM95–YM88
DSP95   = SP95–SP94
DD95   = D95–D94

Hypothesis 3 is that:

> Endogenous growth is more likely to occur in competitive local economies rather than in local economies dominated by one or relatively few large firms.

This hypothesis is tested using models for individual high-tech products. The modeling for the cross-sectional and time-series analyses is basically the same, and thus the same notation is used. The difference is that the operational definitions of the variables need to be adjusted to reflect high-tech activities at the level of individual product. Some variables, such as total high-tech employment growth (TYG95), specialization index (SP), diversity index (D), competition (COMP88), and persistence (Y88) are operationally redefined as in Table 3.3. Also, the manufacturing employment growth variable (TYGOM95) and persistence of manufacturing industries (YOM88) are replaced by total employment growth of other high-tech industries (TYGOHT95) and persistence of other high-tech industries (YOHT88), respectively.

For the cross-sectional analysis, the three equations are rewritten as follows:

$$\text{TYG95} = f\,(\text{SP88 D88 MW88}$$
$$\qquad\qquad \text{COLL ACCESS COMP88 Y88 CMSA MSA}) \qquad\qquad (26)$$
$$\text{SP95} = g\,(\text{TYG95 MW88}$$
$$\qquad\qquad \text{COLL ACCESS COMP88 Y88 CMSA MSA}) \qquad\qquad (27)$$

Table 3.3 Operational Definitions of the Variables:
Cross-sectional Analysis for Individual High-tech Product

Code	Variable	Operational Definition
TYG95	total high-tech employment growth	employment change of high-tech product k in location i between 1988 and 1995
SP88, SP95	specialization index for product k	ratio of employment for product k per total high-tech employment in location i relative to those in the state of Texas for the years 1988 and 1995
D88, D95	diversity index (Hirschman-Herfindahl Index) for product k	Σ (employment share for j)5 $= \Sigma$ (employment for j / total high-tech employment)5 for any other j high-tech product in location i for the years 1988 and 1995

COMP88	local competition for product k	ratio of establishments per worker for product k relative to those in the state of Texas in location i in 1988
Y88	persistence for product k	initial high-tech employment level for product k in location i in 1988
TYGOHT95	total employment growth of other high-tech industries	employment change of high-tech industries in location i between 1988 and 1995
YOHT88	persistence of other high-tech industries	initial level of high-tech industries in location i in 1988

$$D95 = h\,(\text{TYG95 MW88}$$
$$\text{TYGOHT95 YOHT88 COLL ACCESS COMP88 Y88 CMSA MSA)} \quad (28)$$

For the time-series analysis and simultaneous equation model, the three equations are rewritten as follows:

$$YG95 = f\,(\text{DSP95 DSP94 DSP93 DSP92 DSP91 DSP90 DSP89}$$
$$\text{DD95 DD94 DD93 DD92 DD91 DD90 DD89}$$
$$\text{MW95 COLL ACCESS COMP88 CMSA MSA)} \quad (29)$$

Table 3.4 Operational Definitions of the Variables:
Time-series Analysis for Individual High-tech Product

Code	Variable	Operational Definition
YG95	high-tech employment growth	employment change of high-tech product k in location i between 1994 and 1995
DSP95 DSP94 DSP93 DSP92 DSP91 DSP90 DSP89	yearly change of specialization for high-tech product k	DSP95=SP95–SP94 DSP94=SP94–SP93 DSP93=SP93–SP92 DSP92=SP92–SP91 DSP91=SP91–SP90 DSP90=SP90–SP89 DSP89=SP89–SP88
DD95 DD94 DD93 DD92 DD91 DD90 DD89	yearly change of diversity (Hirschman-Herfindahl Index) for high-tech product k	DD95=D95–D94 DD94=D94–D93 DD93=D93–D92 DD92=D92–D91 DD91=D91–D90 DD90=D90–D89 DD89=D89–D88

NOTE: The lagged variables of specialization and diversity are defined as in Table 3.4.

SP95 $= g$ (TYG95

  DMW95 DMW94 DMW93 DMW92 DMW91 DMW90 DMW89

  COLL ACCESS COMP88 Y88 CMSA MSA)   (30)

D95 $= h$ (TYG95

  DMW95 DMW94 DMW93 DMW92 DMW91 DMW90 DMW89

  TYGOHT95 YOHT88 COLL ACCESS COMP88 Y88 CMSA MSA) (31)

There are also three identities:

TYG95 $=$ Y95–Y88

DSP95 $=$ SP95–SP94

DD95 $=$ D95–D94

Table 3.4 summarizes the operational definitions for the variables used for the time-series analysis.

  In equation (26), the coefficient on the competition variable (COMP88) represents a direct effect on total high-tech employment growth (TYG95), since the influence of the past specialization and diversity (SP88 and D88) is held constant when this equation is estimated. According to Jacobs' and Porter's hypothesis, the expected sign should be positive. In equations (27) and (28), the coefficients of the competition variables represent the effects on current specialization and diversity, since the influence of high-tech employment growth is held constant. In both equations, a negative sign of the competition variable is expected. The interpretation for the competition variable (COMP88) in the simultaneous equation model is the same as that of cross-sectional analysis. In equation (29), the direct effect of the competition variable on high-tech employment growth can be estimated. In equations (30) and (31), it is the indirect effect of competition variable on high-tech employment growth that is estimated.

3.4 Data Sources

Two different units of analysis are used to test the three hypotheses: metropolitan areas and counties in non-metropolitan areas,[2] and municipalities. All the models are estimated at these two spatial scales, as the data permit. The data are derived primarily from successive issues of the *Texas High Technology Directory*, yearly reports of County Business Patterns (CBP), and the 1990 Census. Yearly data for high-tech firms in Texas came from the Directory, and I calculated the measures of specialization, diversity (Hirschman-Herfindahl Index), and competition. The specialization index represents MAR externalities directly since it facilitates spillover information flows among relevant firms. The diversity index measures all high-tech economic activity, such as microelectronics, communications equipment, aerospace/aircraft equipment, software activities, etc. (as discussed later). The CBP data are used to construct annual wage rates for manufacturing industries, and annual measures of manufacturing employment. The data pertaining to availability of highly-skilled labor (percentage of the population more than 25 years old with at least college degree) is taken from the 1990 Census.

Regarding high-tech establishments in Texas, the Directory defines high-tech as establishments to develop and/or manufacture proprietary products that incorporate state-of-the-art technology. Exceptions are made with software firms, research, development, and testing companies and laboratories that have significant regional presence and technical expertise. This innovation-oriented definition is more likely to reflect the nature of high-tech firms. In terms of 3-digit SIC codes, Table 3.5 compares this product sophistication definition with the occupation-based definition which Markusen et al. (1986: 18-19) establish at the national level using the 1980 Occupational Employment Statistics (OES). The OES data exclude important service sectors, such as computer software and commercial R&D labs, which deserve the high-tech label. The differences seem to reflect recent transitions of high-tech industry.

Table 3.5 Comparison of High-tech Definitions

	Product Sophistication Definition*	**Occupation-Based Definition****
Commonalities	281 industrial inorganic chemicals 282 plastics, materials and synthetics 283 drugs 284 cleaning preparations 286 industrial organic chemicals 287 agricultural chemicals 289 miscellaneous chemical products 348 ordnance and accessories 351 engines and turbines 353 energy/material handling machinery 354 metalworking machinery 356 general industrial machinery 357 computer and office equipment 361 electrical distribution equipment 362 electrical industrial apparatus 365 household audio and video equipment 366 communications equipment 367 electronic components and accessories 372 aircraft and parts 376 guided missiles, space vehicles, parts 381 search and navigation equipment 382 measuring and controlling instruments 384 lab, medical instruments and supplies 386 photographic equipment and supplies	Categories are the same as those in the product sophistication column.

Differences	173 electrical work (telecommunications)	285 paints and varnishes
	299 miscellaneous chemical products	291 petroleum refining
	308 miscellaneous plastics	303 reclaimed rubber
	322 glass and glassware, pressed or blown	374 railroad equipment
	326 pottery and related products	383 optical instruments
	329 miscellaneous nonmetallic	and lenses
	mineral prod.	
	335 non-ferrous metals	
	339 miscellaneous primary metal prod.	
	344 fabricated structural metal prod.	
	349 miscellaneous fabricated metal prod.	
	355 special industry machinery	
	358 refrigeration and service industry	
	machinery	
	359 miscellaneous industrial machinery	
	364 electric lighting and wiring equipment	
	369 misc. electrical equipment and supplies	
	385 ophthalmic goods	
	399 miscellaneous manufacturing	
	industries	
	737 computer software and services	
	807 medical and dental laboratories	
	873 research, development and testing	
	services	
	899 other services (environmental)	

NOTE: * *Texas High Technology Directory*, Ashland: Leading Edge Communications, Inc., 1995.

    ** Markusen et al., *High Tech America*, 1986, Boston, pp.18–19.

Table 3.6 Distribution of Texas High-tech Establishments across Municipalities and Metropolitan Areas and Non-metropolitan Counties, 1988–1995

	1988	1989	1990	1991	1992	1993	1994	1995
Establishments	1,371	1,802	2,299	2,598	3,092	3,485	3,806	4,001
Municipalities	165	205	252	288	318	324	329	334
MAs and non-metropolitan counties	61	78	89	99	111	112	115	115

Source: *Texas High Technology Directory*, Ashland: Leading Edge Communications, Inc., 1988–1989, 1989–1990, 1990–1991, 1991–1992, 1993, 1994, 1995 and 1996.

From the Directory, I was able to extract eight years of panel data for the period of 1988–1995.[3] The data are composed of 4,001 high-tech firms across 334 municipalities (and 115 metropolitan areas and non-metropolitan counties) in 1995, as shown in Table 3.6. Texas high-tech industry has been developing very rapidly since the 1970s and continues to provide a robust industrial base in the early 1990s. A great many Texas cities have become the home of high-tech industrial sectors (e.g., communications, electronics, aviation, and computers). Also, since the 1970s, state and city governments have actively pursued policies and programs to further the growth of high-technology industries. By using Texas data, I am able to control for social and political factors to some extent. The panel data analyses were constructed to permit exploration of the causal linkages among endogenous variables which in the model are simultaneously determined.

The high-tech industry data are disaggregated across municipalities into 30 high-tech products (26 products for 1994 to 1995). Disaggregation to the product level seems to be necessary to unravel the important high-tech activities at the level of the individual product. Since there might be some similarities among products, I combine them into 13 products for estimation on the basis of major SIC codes. Most of these products also can be clustered into the three major industry ensembles ('intellectual-capital' industries, 'innovation-based' industries, and 'variety-based' industries), proposed by Pollard et al. (1996: 1). Table 3.7 summarizes product classifications in terms of major SIC codes, geographical distribution, and percentage of the major SIC codes covered.

Table 3.7 Product Classification

1. Innovation-based Industries

Product	Contents	Major SIC codes	1988 firms (munici-palities)	1995 firms (munici-palities)	% of Major SIC codes
1) micro-electronics	electronic components and assembly and related research	367, 873	195 (44)	405 (93)	90.1
2) computer systems (hardware) and peripherals/ accessories	electronic computers, computer peripherals/ accessories, storage peripherals, computer graphics, and related research	357, 873	127 (27)	279 (52)	86.5

3) medical equipment/ devices	medicial electronics/ instruments and supplies	384	38 (18)	164 (62)	92.0
4) aerospace/aircraft and equipment	aircraft and parts, guided missiles, space vehicles and parts	372, 376	49 (25)	221 (75)	92.5
5) military equip- ment/services	ordnance, air- craft and parts, guided missiles, search and navigation equipment and related research	348, 372, 376, 381, 873	36 (16)	85 (42)	92.0

NOTE 1: The percentage of major SIC codes is calculated for each product, using *Texas High Tech Directory*, Ashland: Leading Edge Communications, Inc., 1995.

NOTE 2: Medical electronics is also classified into innovation-based industry that is defined as those employing high-proportions of highly skilled, technical labor.

Table 3.7 (continued)

2. Intellectual-capital Industries

Product	Contents	Major SIC codes	1988 firms (municipalities)	1995 firms (municipalities)	% of Major SIC codes
1) communications equipment/services	telecommunication, communications equipment, broadcast equipment, and video	173, 365, 366	127 (39)	298 (83)	91.7
2) measurement and testing equipment	analytical/ measuring instruments and test equipment, monitoring/ controlling devices and electrical industrial apparatus	382, 362	336 (79)	599 (142)	90.2
3) software development/ services	computer software and services	737	241 (36)	779 (86)	99.3
4) research, development and testing	research, development and testing services	873 (12)	40 (60)	194	94.4

NOTE 1: The percentage of major SIC codes is calculated for each product, using *Texas High Tech Directory*, Ashland: Leading Edge Communications, Inc., 1995.

NOTE 2: Research, development and testing is also classified into intellectual capital industry that is defined as those with high proportions of high-wage, non-production occupations.

Table 3.7 (continued)

3. Variety-based Industries

Product	Contents	Major SIC codes	1988 firms (munici-palities)	1995 firms (munici-palities)	% of Major SIC codes
1) chemicals	inorganic and organic chemicals, agricultural chemicals and synthetics	281, 282, 284, 286, 287, 289, 299	35 (14)	492 (137)	95.2
2) pharmaceuticals and biological products	biological products and related research, drugs and related research	283, 873	47 (22)	108 (50)	90.5
3) material handling equipment	construction and related machinery	353	13 (10)	117 (52)	93.6
4) other high-tech manufacturing	components, plastics/advanced materials, lasers/ optics/photonics, electronics production equipment, industrial equipment robotics/ factory automation, energy, power devices/ systems, non-industrial electrical products, and related research and services (environmental)	282, 308, 326, 329, 335, 344, 351, 353, 354, 355, 356, 357, 358, 359, 361, 364, 366, 367, 369, 382, 399, 873, 899	399 (83)	1,230 (208)	87.2

NOTE 1: The percentage of major SIC codes is calculated for each product, using *Texas High Tech Directory*, Ashland: Leading Edge Communications, Inc., 1995.

NOTE 2: Other high-tech manufacturing industries are classified into variety-based industry that is defined on the basis of product diversification, although some of them are either innovative-based industry or intellectual capital industry.

3.5 Models of the Effects of High-tech Development on Regional Employment Growth

Finally, I need to address the policy issues that I raised in chapter 2. The main issue was, first, whether there is a positive relationship between high-tech development and regional economic growth. A related issue was to assess the validity of the two general approaches to promoting high-technology economic development – attracting high-tech firms or encouraging regional business initiative. By using annual data for high-tech employment growth in Texas for the period of 1988 to 1995, the yearly trend can be compared with employed civilian labor force (CLF) in the same period. Based on product classification, firm size or organizational type, the growth patterns of high-tech industry also can be discussed.

The first step is to estimate a cross-sectional model of the effects of high-tech development on regional economic growth. The following model hypothesizes that initial high-tech level is positively associated with employed CLF growth. Nonlinear models are used to describe the trend. The growth of employed CLF in location i is modeled as a function of initial level of high-tech employment in that location and other control variables representing regional characteristics as follows:

$$LNECLF950 = f(Y90\ YSQ\ MANU\ CMSA\ MSA) \tag{32}$$
$$LNECLF950 = g(Y90\ YRV\ MANU\ CMSA\ MSA) \tag{33}$$

where LNECLF950 = log of employment change in employed CLF between 1990 and 1995 in location i,

Y90 = log of initial high-tech employment level in 1990 in location i,
YSQ, YRV = the squared and inverse terms of Y90,
MANU = percentage of manufacturing industries in 1990 in location i,
CMSA, MSA = dummies for the presence in CMSA or MSA.

I also examine the differences in the effects of initial employment level of single firms and multi-locational establishments on employed CLF growth. The model is expanded as follows:

$$LNECLF950 = f(YSI90\ YM90\ YSQ\ MANU\ CMSA\ MSA) \tag{34}$$
$$LNECLF950 = g(YSIRV\ MANU\ CMSA\ MSA) \tag{35}$$
$$LNECLF950 = h(YMRV\ MANU\ CMSA\ MSA) \tag{36}$$

where YSI90, YM90 = log of initial employment level of single firms and multi-locational establishments, respectively, in 1990 in location i,

YSIRV, YMRV = the inverse of initial employment level of single firms and multi-locational establishments, respectively, in 1990 in location i.

These models are estimated by OLS, using cross-sectional data for the years 1990–1995, when the employed CLF data is available. The CLF data for metropolitan areas and non-metropolitan counties came from the reports from the

Bureau of Labor Statistics, whereas the CLF data for municipalities came from the Texas Workforce Commission. The percentage of manufacturing industries is operationally defined as the percentage of manufacturing employment per employed CLF in 1990 in location i.

Notes

1. In this study, I use municipalities as geographical units.
2. The metropolitan areas are composed of Consolidated MSAs (CMSAs) and MSAs. In this study, all legal boundaries and names are as of January 1, 1990, as defined by the U.S. Bureau of the Census. The list of the metropolitan areas and countries involved is provided in Appendix B.
3. The definition is published at the beginning of the year.

Chapter 4

Empirical Findings

I now report the empirical results for all the models that were estimated. First, I discuss the regression results regarding the regional high-tech employment effects of dynamic externalities (hypothesis 1). Next, I consider the differences in these impacts by firm size, organizational type, and product (hypotheses 2 and 3). Finally, the regression results regarding the effects of initial high-tech employment on regional growth of employed civilian labor force are discussed.

4.1 Regional High-tech Employment Effects of Dynamic Externalities

In this section, I report on the cross-sectional and simultaneous equation tests of hypothesis 1, namely that regional growth in high-tech employment is determined by endogenous growth processes. First, I discuss in sections 4.1.1 and 4.1.2 the solutions to equations (4), (5), and (6). Then I discuss the solutions to equations (7), (8), and (9) in sections 4.1.3 and 4.1.4.

4.1.1 The Effects of Historical Conditions on High-tech Employment Growth: Cross-sectional Analysis

The first question to be addressed is that of the effects of historical conditions on high-tech employment growth. Equation (4) states that high-tech employment growth (TYG95) is a function of past specialization (SP88), past diversity (D88), persistence (initial level) of high-tech employment (Y88), controlling for manufacturing wage (MW88), availability of highly-skilled labor (COLL), access to international and national markets (ACCESS), local competition (COMP88), and the dummies for CMSA and MSA. In this equation, historical conditions are described by three measures: two relating to own industry employment in 1988 (SP88 and Y88), and one relating to industrial diversity in 1988 (D88).

The regression results are laid out in Table 4.1. The second column reports the coefficients when metropolitan areas and non-metropolitan counties are the units of observation. The fifth column reports the coefficients for municipalities as the units of observation. There are several important findings:

1. The degree of past specialization (SP88) strongly affected regional high-tech employment growth between 1988 and 1995 (TYG95) in Texan metropolitan areas, but when municipalities are used as the units of observation, the effects of specialization are not readily observable: the higher growth rates are in municipalities newly acquiring their high-tech foundations. High-tech concentrations grew rapidly in large metropolitan regions. The coefficient for the

Table 4.1 The Effects of Historical Conditions on High-tech Employment Growth: Cross-sectional Analysis: Total Products

DEP. VARIABLE: HIGH-TECH EMPLOYMENT GROWTH (TYG95)

VAR.	MAs & COUNTIES			MUNICIPALITIES		
	COEFFICIENTS	T STAT.	PROB> |T|	COEFFICIENTS	T STAT.	PROB> |T|
INTERCEP	−4873.642298*	−1.812	0.073	−376.55719	−0.184	0.8541
SP88	5378.855158**	4.838	0.0001	554.645208	0.584	0.5597
D88	249.014943	0.112	0.9109	−741.696338	−0.515	0.6066
Y88	−0.204883**	−4.179	0.0001	0.986641**	10.282	0.0001
MW88	0.107242	1.586	0.1158	0.04001	0.816	0.4151
COLL ʹ	115.21833	1.082	0.2819	−0.004325	0	0.9999
ACCESS	3.665948	0.975	0.3318	1.884543	0.514	0.6075
COMP88	5.681641	0.338	0.7362	1.684733	0.104	0.9169
CMSA	123721**	21.047	0.0001	−151.890343	−0.155	0.877
MSA	1978.125924	1.404	0.1633	−51.876031	−0.063	0.9502
F-VALUE		170.74			37.779	
R-SQ		0.9389			0.5313	
ADJ. R-SQ		0.9334			0.5172	
N		110			310	
CON. INDEX		19.0159			18.8515	

NOTE: * = significance at the .10 level, ** = significance at the .05 level.

specialization variable is statistically significant and quantitatively large. The standard deviation of SP88 was 0.4635. The MAR effect measured by high-tech employment growth of a one-standard deviation increase in past specialization was 2,493 employees. This indicates that historical specialization of the high-tech industry creates an environment conducive to attracting high-tech industry. The results strongly support the importance of within-industry knowledge spillovers for growth. The findings are consistent with Henderson et al. (1995) and Miracky (1992), although the samples and time periods differ, and differ from those in Glaeser et al. (1992). Glaeser et al. include an industry only if it is one of the six largest industries in a city. Non-traded sectors are heavily represented in their sample. As Henderson et al. (1995) argue, within-industry knowledge spillovers may not matter for mature industries.

2. Historical diversity has no significant impact on subsequent high-tech employment growth: Jacobs effects are not observed.
3. Consistent with Verdoorn's law, high initial employment in high-tech industry leads to rapid growth of this industry, but this generalization is only true for municipalities. For metropolitan areas, the variable has the opposite sign: the higher the initial metropolitan agglomeration, the lower the high-tech growth. The most rapid high-tech growth is in municipalities outside metropolitan regions as high-tech employment diffuses across the state.

4. Among the control variables, initial manufacturing wage (MW88) has positive sign, although not statistically significant. Manufacturing wage could reflect regional differences in technical expertise. This is consistent with Glaeser et al. (1994), arguing that the urban wage premium accrues over time as a result of greater skill accumulation in cities. The human capital variable (COLL) also has positive effects on subsequent employment growth in metropolitan areas. However, these variables are less significant for municipalities, due mainly to the spread-out phenomenon of high-tech industry. As expected, metropolitan areas grew significantly more than non-metropolitan counties, but for municipalities, the presence in a CMSA or MSA affected high-tech employment growth negatively. Other control variables such as the access variable (ACCESS) and the initial competition variable (COMP88) are not statistically significant.

4.1.2 The Determinants of the Degree of Dynamic Externalities: Cross-sectional Analysis

The second question to be addressed relates to determinants of the degree of dynamic externalities. Equations (5) and (6) are relevant here. Equation (5) states that current specialization (SP95) is a function of high-tech employment growth (TYG95), persistence (initial level) of high-tech employment (Y88), controlling for manufacturing wage (MW88), availability of highly-skilled labor (COLL), access to international and national markets (ACCESS), local competition (COMP88), and the dummies for CMSA and MSA.

The regression results are laid out in Table 4.2. Major findings are as follows:

1. Current specialization (SP95) is highly dependent upon total high-tech employment growth (TYG95) and initial high-tech level (Y88). The high-tech employment variables are positive and significant at both spatial scales. A higher level of initial high-tech employment leads to more specialized high-tech locations.

   Interestingly, CMSAs became less specialized than MSAs and non-metropolitan counties over the period of 1988 to 1995. This relationship is attributed to the existence of agglomeration economies in the first stages of development and external diseconomies at more advanced levels of development. In effect, there is strong evidence for the bell-shaped relationship between urban concentration and economic development (Petrakos, 1992: 1221). This is consistent with many historical cases such as Silicon Valley and Route 128 surrounding Boston.

Table 4.2 The Determinants of the Degree of Specialization: Cross-sectional Analysis: Total Products

DEP. VARIABLE: SPECIALIZATION (SP95)

VAR.	MAs & COUNTIES			MUNICIPALITIES		
	COEFFICIENTS	T STAT.	PROB> |T|	COEFFICIENTS	T STAT.	PROB> |T|
INTERCEP	−0.256561*	−1.72	0.0886	−0.066802	−0.988	0.3239
TYG95	0.028726**	4.254	0.0001	0.034658**	12.021	0.0001
Y88	0.008949**	2.515	0.0135	0.035424**	8.46	0.0001
MW88	0.016573**	3.179	0.002	0.007125**	2.903	0.004
COLL	0.008555	1.076	0.2844	0.004403**	2.729	0.0067
ACCESS	0.000761**	2.643	0.0095	0.000502**	2.747	0.0064
COMP88	−0.001087	−0.866	0.3886	−0.000402	−0.504	0.6148
CMSA	−3.285744**	−3.684	0.0004	−0.213841**	−4.512	0.0001
MSA	−0.033996	−0.375	0.7083	−0.091614**	−2.238	0.026
F-VALUE		10.655			103.046	
R-SQ		0.4577			0.7325	
ADJ. R-SQ		0.4147			0.7254	
N		110			310	
CON. INDEX		13.9217			12.5934	

NOTE 1: * = significance at the .10 level, ** = significance at the .05 level t statistics are in parentheses.

NOTE 2: The coefficients for TYG95, Y88, and MW88 are multiplied by 1,000 persons.

2. Other control variables have the expected signs. Initial wage (MW88) and availability of highly skilled labor (COLL) are positively associated with current specialization (SP95). The effects of these variables are highly significant: the greater the skilled work force, the greater the opportunity for high-tech industry to benefit from economies of scale. The sign of the access variable (ACCESS) is positive and statistically significant, reflecting the diffusion of high-tech industry. Competition variable (COMP88) has the expected sign, but is insignificant.

Next question is the significance of high-tech employment as the determinant of the degree of diversity. Equation (6) states that current diversity (D95) is a function of high-tech employment growth (TYG95), persistence (initial level) of high-tech employment (Y88), controlling for regional characteristics such as manufacturing wage (MW88), total employment growth of other manufacturing industries (TYGOM95), initial level of other manufacturing industries (YOM88), availability of highly-skilled labor (COLL), access to international and national markets (ACCESS), local competition (COMP88), and the dummies for CMSA and MSA. In this equation, current diversity for high-tech industry is assumed to depend on the initial employment level and growth of manufacturing industries, as well as on those of high-tech industry.

Table 4.3 summarizes the regression results. The major findings are as follows:

1. Current diversity (D95) is highly dependent on total high-tech employment growth (TYG95) and initial high-tech level (Y88). Higher initial employment levels created more diversified environments. High-tech growth also was negatively significant for metropolitan areas. CMSA and MSA were more diversified than elsewhere.
2. Current diversity was affected by the control variables in various ways. As expected, employment growth of other manufacturing industries (TYGOM95) was one factor to affect current diversity of high-tech industry. High portions of highly-skilled labor (COLL) created more diversified environments for high-tech industry. More interestingly, the competition variable (COMP88) was negatively significant. One possible interpretation is that on average more firms per worker lead to more diversified environments. Another interpretation of this finding is that smaller firms created more diversified environments. This is consistent with Jacobs' or Porter's views.

In summary, the cross-sectional analysis suggests that there is simultaneity between high-tech employment growth and dynamic externalities. The next section reports the empirical results concerning the simultaneous model.

4.1.3 The Effects of Historical Conditions on High-tech Employment Growth: Time-series Analysis

This section reports the empirical results for the simultaneous equation model for total products. Current employment growth (YG95) is assumed to be simultaneously determined with current dynamic externalities (SP95 and D95) and current manufacturing wage (MW95). Equations (7), (8), and (9) are relevant here.

Equation (7) states that current high-tech employment growth (YG95) is a function of yearly differences of specialization (DSP), yearly differences of diversity (DD), persistence (initial level) of high-tech employment (Y88), controlling for manufacturing wage (MW95), availability of highly-skilled labor (COLL), access to international and national markets (ACCESS), local competition (COMP88), and the

Table 4.3 The Determinants of the Degree of Diversity:
Cross-sectional Analysis: Total Products

DEP. VARIABLE: DIVERSITY (D95)

VAR.	MAs & COUNTIES			MUNICIPALITIES		
	COEFFICIENTS	T STAT.	PROB> |T|	COEFFICIENTS	T STAT.	PROB> |T|
INTERCEP	0.849545**	7.168	0.0001	0.949487**	13.854	0.0001
TYG95	−0.025552**	−2.741	0.0073	0.002059	0.528	0.5979
Y88	−0.006551*	−1.7	0.0923	−0.017399**	−4.09	0.0001
MW88	−0.002412	−0.573	0.5681	−0.002288	−0.918	0.3593
TYGOM95	−0.060841**	−2.935	0.0041	−0.002483	−0.822	0.4118

YOM88	−0.009263	−1.592	0.1146	0.000578	1.501	0.1345
COLL	−0.007039	−1.109	0.2703	−0.00739**	−4.503	0.0001
ACCESS	0.000201	0.894	0.3733	−0.000048583	−0.263	0.7928
COMP88	−0.001655*	−1.675	0.0971	−0.001137	−1.401	0.1621
CMSA	−1.91846	−1.426	0.1571	−0.135291**	−2.618	0.0093
MSA	−0.191252**	−2.191	0.0308	−0.069085*	−1.656	0.0988
F-VALUE		5.239			8.845	
R-SQ		0.3461			0.2283	
ADJ. R-SQ		0.28			0.2025	
N		110			310	
CON. INDEX		35.8504			13.2065	

NOTE 1: * = significance at the .10 level, ** = significance at the .05 level t statistics are in parentheses.

NOTE 2: The coefficients for TYG95, Y88, MW88, TYGOM95, and YOM88 are multiplied by 1,000 persons.

dummies for CMSA and MSA. The regression results for equation (7) are laid out in Table 4.4. Since the Hausman specification test shows simultaneity between the fitted values from the model and the error terms, the results reported are two-stage least-squares estimates. The major findings are as follows:

1. Yearly differences of specialization have positively significant effects on current high-tech growth. The evidence supports the MAR view. Figure 4.1 clearly shows the lagged effects of specialization for current high-tech employment. As Henderson (1994) argues, lagged specialization variables have about 5–6 years of time-lag. This is relatively little, because one would expect to have a much longer historical impact on current employment. For municipalities, the lagged effects seem to be somewhat longer.

Table 4.4 The Effects of Historical Conditions on High-tech Employment Growth: Time-series Analysis: Total Products Two-Stage Least Squares Estimates

DEP. VARIABLE: HIGH-TECH EMPLOYMENT GROWTH (YG95)

	MAs & COUNTIES			MUNICIPALITIES		
VAR.	COEFFICIENTS	T STAT.	PROB> |T|	COEFFICIENTS	T STAT.	PROB> |T|
INTERCEP	−1140.44	−0.812	0.4196	517.6543	1.092	0.276
DSP95	14594**	2.142	0.0355	15321**	3.973	0.0001
DSP94	1634.563	0.691	0.4917	1309.131*	1.694	0.0913
DSP93	3009.274	1.43	0.1569	1430.576**	2.606	0.0097
DSP92	3995.811	1.607	0.1122	2314.782**	2.931	0.0037
DSP91	2604.876	1.276	0.2061	4845.387**	4.077	0.0001
DSP90	−716.458	−0.645	0.5207	3114.473**	3.676	0.0003

DSP89	−866.148	−1.002	0.3197	862.8004**	2.148	0.0326
DD95	9950.977	1.298	0.1984	1413.028	0.364	0.7158
DD94	5322.727*	1.691	0.095	1471.342	0.8	0.4245
DD93	2231.139	1.086	0.2811	1653.506	1.22	0.2236
DD92	2135.541	0.836	0.4058	568.5633	0.565	0.5725
DD91	3296.276	1.483	0.1423	125.8052	0.142	0.8873
DD90	2726.24	1.608	0.1121	807.889	0.951	0.3422
DD89	1149.272	0.748	0.4568	1149.349*	1.864	0.0635
Y88	0.1285**	4.501	0.0001	0.315703**	13.209	0.0001
MW95	0.048333	1.29	0.2011	−0.00482	−0.349	0.7277
COLL	131.5524*	1.793	0.077	7.362765	0.677	0.4991
ACCESS	−1.715	−0.556	0.5802	−0.6131	−0.474	0.6358
COMP88	4.864164	0.277	0.7827	−3.23071	−0.571	0.5683
CMSA	8422.785**	2.225	0.0291	−419.387	−1.391	0.1653
MSA	2.805782	0.003	0.9972	36.10908	0.136	0.8919
F-VALUE		13.842			23.618	
R-SQ		0.7971			0.645	
ADJ. R-SQ		0.7395			0.6177	
N		96			295	

NOTE: * = significance at the .10 level, ** − significance at the .05 level.

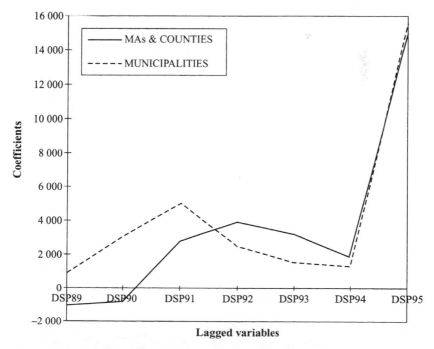

Figure 4.1 The Lagged Effects of Specialization on High-tech Employment Growth: Total Products

2. Unexpectedly, the lagged diversity variables have positive signs: the lack of diversity is positively associated with high-tech employment growth. As in cross-sectional analysis, Jacobs effects are not observed.
3. A very high degree of persistence was observed across metropolitan areas and municipalities, despite strong evidence for relocation. Part of the glue that holds concentrations in specific locations overtime is thought to be the remaining effects of MAR externalities (Henderson et al., 1995). Locations with historical concentrations of high-tech industry can offer a more productive environment than those without them. Other things being equal, CMSAs gained high-tech employment for the period of 1988–1995.
4. Availability of highly-skilled labor (COLL) was positively significant. This confirms the importance of human capital in high-tech industry. Other control variables were not statistically significant.

4.1.4 The Determinants of the Degree of Dynamic Externalities: Time-series Analysis

This section deals with equations (8) and (9). As in cross-sectional analysis, current specialization (SP95) and current diversity (D95) are assumed to depend on total high-tech employment (TYG95) and persistence (initial level) of high-tech employment (Y88), controlling for regional variables and dummies. Unlike the cross-sectional analysis, yearly differences of manufacturing wages (DMW) are included as control variables. To reduce the multicollinearity problem, first differences are used.

Table 4.5 shows the two-stage least-squares estimates regarding the impact of high-tech initial level and growth on specialization. The major findings are as follows:

1. Current specialization (SP95) is highly dependent upon total high-tech employment growth (TYG95) and initial high-tech level (Y88) in both spatial scales. A higher initial level of high-tech employment and subsequent growth lead to more specialized high-tech locations. But, CMSAs became less specialized than elsewhere. All the results confirm the findings in the cross-sectional analysis.
2. Wage effects seem to persist for four or five years as shown in Figure 4.2. For this period, wage differentials have negative signs: wage increases reduced the concentrations of high-tech industry. A number of years are statistically significant. This result is consistent with Henderson et al. (1994), although they used logs in employment as the dependent variable. Wages can be viewed as affecting firm factor proportions, with rapid adjustment.
3. A high degree of availability of highly-skilled labor (COLL) leads to more specialized locations. This effect was significant especially for municipalities. The access variable has positive signs, as in the cross-sectional analysis, reflecting diffusion of high-tech industry.

Table 4.6 shows the two-stage least-squares estimates regarding the impact of initial level and growth of high-tech industry on diversity. The major findings are as follows:

1. Current diversity (D95) was negatively associated with total high-tech employment growth (TYG95) and initial high-tech employment. Higher levels of initial high-tech employment lead to significantly diversified environments. High-tech locations tend to attract supporting industries as well as their own industries. This idea is consistent with successful local economic development policies. The signs of the CMSA and MSA dummies were all negative.
2. Wage effects seem to persist for three or four years as shown in Figure 4.2. For this period, the signs of lagged differential wage variables were positive, meaning that wage increases reduced diversity. For a shorter period, other things being equal, the wage variable seems to be negatively associated with dynamic externalities.

Table 4.5 The Determinants of the Degree of Specialization:
Time-series Analysis: Total Products
Two-Stage Least Squares Estimates

DEP. VARIABLE: SPECIALIZATION (SP95)

	MAs & COUNTIES			MUNICIPALITIES		
VAR.	COEFFICIENTS	T STAT.	PROB> |T|	COEFFICIENTS	T STAT	PROB> |T|
INTERCEP	0.054339	0.348	0.7286	0.114721*	1.837	0.0672
TYG95	0.036405**	4.366	0.0001	0.03967**	10.608	0.0001
Y88	0.009596**	2.506	0.0142	0.030657**	6.263	0.0001
DMW95	0.000841	0.074	0.9416	−0.00253	−0.558	0.5775
DMW94	−0.016681	−0.792	0.4305	−0.028692**	−2.521	0.0122
DMW93	−0.019459	−0.865	0.3897	−0.024853*	−1.834	0.0677
DMW92	−0.006899	−0.411	0.6824	−0.00485	−0.52	0.6038
DMW91	0.004942	0.271	0.7867	−0.00273	−0.308	0.7586
DMW90	0.030134	1.375	0.1728	0.013197	1.082	0.2803
DMW89	0.036382	1.568	0.1208	0.005779	0.551	0.582
COLL	0.007425	0.838	0.4044	0.00462**	2.806	0.0054
ACCESS	0.000801**	2.266	0.0261	0.00037*	1.837	0.0672
COMP88	−0.000922	−0.598	0.5512	−0.000568	−0.698	0.4855
CMSA	−3.990824**	−3.701	0.0004	−0.147247**	−3.291	0.0011
MSA	−0.020396	−0.212	0.8328	−0.050742	−1.245	0.2141
F-VALUE		5.289			55.972	
R-SQ		0.4776			0.7367	
ADJ. R-SQ		0.3873			0.7236	
N		96			295	

NOTE 1: * = significance at the .10 level, ** = significance at the .05 level.
NOTE 2: The coefficients for TYG95, Y88, and DMW are multiplied by 1,000 persons.

3. Other important sources of diversity were the employment growth of manufacturing industries (TYGOM95), availability of highly-skilled labor (COLL), and the initial competition (COMP88). These results confirm the findings of the cross-sectional analysis. As expected, all these variables had negative signs.

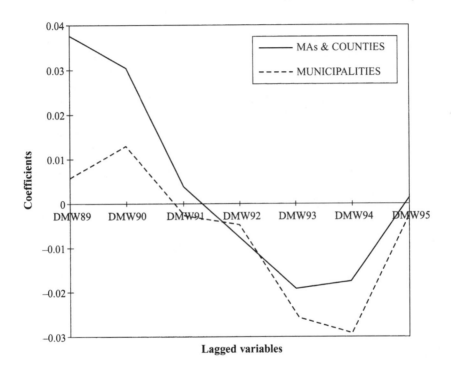

Figure 4.2 The Lagged Effects of Manufacturing Wages on Current Specialization: Total Products

4.2 The Differences in Regional High-tech Employment Effects of Dynamic Externalities by Firm Size

I now turn to the cross-sectional and simultaneous equation tests of hypothesis 2, namely that the growth of small firms is more likely to be driven by endogenous processes than is the growth of large firms. First, I discuss in sections 4.2.1 and 4.2.2 the solutions to equations (10), (11), (12) and (13). Then I discuss the solutions to equations (14), (15), (16) and (17) in sections 4.2.3 and 4.2.4.

Table 4.6 The Determinants of the Degree of Diversity: Time-series Analysis: Total Products Two-Stage Least Squares Estimates

DEP. VARIABLE: DIVERSITY (D95)

VAR.	MAs & COUNTIES			MUNICIPALITIES		
	COEFFICIENTS	T STAT.	PROB> |T|	COEFFICIENTS	T STAT.	PROB> |T|
INTERCEP	0.791636**	6.801	0.0001	0.844077**	12.73	0.0001
TYG95	−0.023748	−1.221	0.2258	−0.003981	−0.413	0.6798
Y88	−0.006473	−1.253	0.214	−0.015051**	−2.624	0.0092
DMW95	−0.009513	−1.122	0.2652	0.001779	0.364	0.7164
DMW94	0.024015	1.552	0.1247	0.017928	1.435	0.1523
DMW93	0.028437*	1.688	0.0953	0.023523*	1.681	0.0938
DMW92	−0.008353	−0.677	0.5006	−0.007457	−0.783	0.4342
DMW91	−0.015216	−1.125	0.264	0.002643	0.268	0.7888
DMW90	−0.011155	−0.694	0.4899	0.00363	0.26	0.7949
DMW89	0.00551	0.325	0.7462	−0.000627	−0.059	0.9531
TYGOM95	−0.055152	−1.47	0.1454	−0.005831	−0.973	0.3314
YOM88	−0.008166	−1.214	0.2283	0.000406	0.987	0.3247
COLL	−0.005073	−0.729	0.468	−0.006881**	−4.085	0.0001
ACCESS	0.000085998	0.326	0.7451	0.000020516	0.097	0.9228
COMP88	−0.00176	−1.458	0.1488	−0.001172	−1.395	0.164
CMSA	−1.737621	−1.173	0.2443	−0.137802**	−2.609	0.0096
MSA	−0.187258**	−2.139	0.0355	−0.063775	−1.453	0.1474
F-VALUE		2.695			5.066	
R-SQ		0.3531			0.2257	
ADJ. R-SQ		0.2221			0.1812	
N		96			295	

NOTE 1: * = significance at the .10 level, ** = significance at the .05 level.
NOTE 2: The coefficients for TYG95, Y88, DMW, TYGOM95, and YOM88 are multiplied by 1,000 persons.

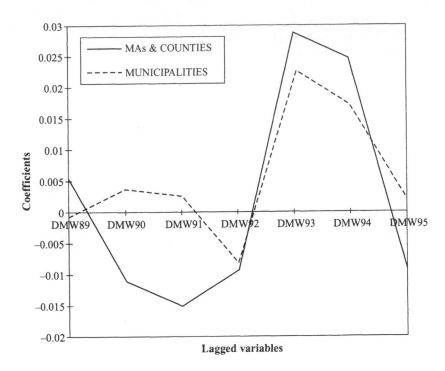

Figure 4.3 The Lagged Effects of Manufacturing Wages on Current Diversity: Total Products

4.2.1 The Effects of Historical Conditions on High-tech Employment Growth: Cross-sectional Analysis

The first question is addressed by estimating equations (10) and (11). In both equations, high-tech employment growth of small and large firms (TYGSM95 and TYGL95), respectively, is a function of past specialization (SP88), past diversity (D88), and persistence (initial level) of high-tech employment (Y88), controlling for regional variables and dummies, which are included in estimating total products.

The regression results are laid out in Tables 4.7 and 4.8. Table 4.7 reports the coefficients when metropolitan areas and non-metropolitan counties are the units of observation. Table 4.8 reports the coefficients for municipalities as the units of observation. There are several important findings:

1. The degree of past specialization (SP88) strongly affects both size of employment growth (TYGSM95 and TYGL95) in Texan metropolitan areas. The coefficients are positively significant and quantitatively large. The MAR effects in large firms are larger than those in small firms. When municipalities are used as the units of observation, however, the effects of specialization for large firms are not readily observable, and the effects for small firms are even negatively significant.

Table 4.7 The Effects of Historical Conditions on High-tech Employment Growth: Cross-sectional Analysis: Small vs. Large Firms MAs and Counties

DEP. VARIABLE: HIGH-TECH EMPLOYMENT GROWTH (TYGSM95 and TYGL95)

VAR.	SMALL COEFFICIENTS	T STAT.	PROB> |T|	LARGE COEFFICIENTS	T STAT.	PROB> |T|
INTERCEP	−106.469218	−0.785	0.4342	−4767.17308*	−1.842	0.0684
SP88	169.287267**	3.021	0.0032	5209.567891**	4.871	0.0001
D88	−63.464879	−0.567	0.5718	312.479821	0.146	0.8839
Y88	0.027037**	10.942	0.0001	−0.23192**	−4.918	0.0001
MW88	0.00292	0.857	0.3934	0.104322	1.604	0.1118
COLL	8.086381	1.506	0.1351	107.131948	1.046	0.2983
ACCESS	0.051518	0.272	0.7862	3.61443	1	0.3199
COMP88	0.018834	0.022	0.9823	5.662807	0.35	0.7271
CMSA	7366.00457**	24.862	0.0001	116355**	20.576	0.0001
MSA	125.074156*	1.762	0.0812	1853.051768	1.367	0.1746
F-VALUE		610.322			152.179	
R-SQ		0.9821			0.932	
ADJ. R-SQ		0.9805			0.9258	
N		110			110	
CON. INDEX		19.0159			19.0159	

NOTE: * = significance at the .10 level, ** = significance at the .05 level.

Table 4.8 The Effects of Historical Conditions on High-tech Employment Growth: Cross-sectional Analysis: Small vs. Large Firms Municipalities

DEP. VARIABLE: HIGH-TECH EMPLOYMENT GROWTH (TYGSM95 and TYGL95)

VAR.	SMALL COEFFICIENTS	T STAT.	PROB> |T|	LARGE COEFFICIENTS	T STAT.	PROB> |T|
INTERCEP	160.210989*	1.746	0.0818	−536.768179	−0.272	0.7859
SP88	−89.909278**	−2.11	0.0357	644.554486	0.703	0.4825
D88	−160.686845**	−2.489	0.0133	−581.009493	−0.418	0.676
Y88	0.076569**	17.788	0.0001	0.910072**	9.827	0.0001
MW88	0.000179	0.082	0.9351	0.039831	0.842	0.4006
COLL	1.286586	0.867	0.3867	−1.290911	−0.04	0.9678
ACCESS	−0.019266	−0.117	0.9068	1.90381	0.538	0.5908
COMP88	−0.394807	−0.545	0.586	2.07954	0.134	0.8939
CMSA	−3.348047	−0.076	0.9394	−148.542295	−0.157	0.8753
MSA	11.130989	0.299	0.765	−63.00702	−0.079	0.9373
F-VALUE		91.493			35.147	
R-SQ		0.733			0.5132	

ADJ. R-SQ	0.725	0.4986
N	310	310
CON. INDEX	18.8515	18.8515

NOTE: * = significance at the .10 level, ** = significance at the .05 level.

2. The degree of past diversity (D88) also strongly affected high-tech employment growth of small firms (TYGSM95) in municipalities. This is in favor of Jacobs' view: Jacobs effects were significantly large. In effect, small firms grew rapidly in both CMSAs and MSAs, whereas large firms grew in CMSAs.
3. Initial high-tech level (Y88) was positively significant for small firms at both spatial scales, but negatively significant for large firms, especially in metropolitan areas. The larger firms as a whole grew slower in heavily concentrated metropolitan areas.

4.2.2 The Determinants of the Degree of Dynamic Externalities: Cross-sectional Analysis

The next question deals with the differences in the determinants of the degree of dynamic externalities by firm size. I focus on the effects of high-tech employment growth of small and large firms on the degree of dynamic externalities. Equations (12) and (13) are relevant here. In both equations, current dynamic externalities (either specialization or diversity) are a function of employment growth of small and large firms (TYGSM95 and TYGL95, respectively), and persistence (initial level) of high-tech employment (Y88), controlling for regional variables and dummies.

The regression results are shown in Table 4.9. The major findings are as follows:

1. Current specialization was affected mainly by employment growth of large firms. The coefficients for TYGL95 was positively significant in metropolitan areas and municipalities. In contrast, the effects of employment growth of small firms was insignificant in metropolitan areas and negatively significant in municipalities. These findings mean that small firms did not play an important role in concentrations of high-tech industry.
2. By grouping high-tech employment by firm size, some control variables become more significant. The manufacturing wage variable (MW88) are positively significant in metropolitan areas and municipalities. The human capital variable (COLL) also is positively significant. Both CMSAs and MSAs are negatively associated with concentrations of high-tech industry for the period of 1988 to 1995.

Table 4.10 shows the regression results regarding the determinants of the degree of diversity. The major findings as follows:

1. Employment growth of small firms (TYGSM95) is significantly correlated with diversity. For both levels of observation, employment growth of small firms had negative effects on current diversity. Employment growth of small firms led to more diversified environments. In contrast, when municipalities are used as the

units of observations, employment growth of large firms shows the opposite sign: it leads to a lack of diversity. The effects of this variable are positively significant.

2. Other sources of diversity were employment growth of manufacturing industries (TYGOM95), availability of highly-skilled labor (COLL), and local competition (COMP88). These variables were negatively significant. The initial level of manufacturing industries was positively significant: a higher initial level deters diversity in municipalities. As expected, CMSAs became more diversified than elsewhere.

Table 4.9 The Determinants of the Degree of Specialization: Cross-sectional Analysis: Small vs. Large Firms

DEP. VARIABLE: SPECIALIZATION (SP95)

	MAs & COUNTIES			MUNICIPALITIES		
VAR.	COEFFICIENTS	T STAT.	PROB> |T|	COEFFICIENTS	T STAT.	PROB> |T|
INTERCEP	−0.256038*	−1.707	0.0909	−0.058488	−0.888	0.3754
TYGSM95	0.001722	0.008	0.9939	−0.357**	−3.78	0.0002
TYGL95	0.029781**	2.675	0.0087	0.048979**	11.006	0.0001
Y88	0.009913	1.124	0.2638	0.049613**	9.32	0.0001
MW88	0.016536**	3.151	0.0021	0.006641**	2.775	0.0059
COLL	0.008636	1.077	0.284	0.005041**	3.192	0.0016
ACCESS	0.000758**	2.615	0.0103	0.000484**	2.718	0.007
COMP88	−0.001089	−0.863	0.3902	−0.000484	−0.62	0.5348
CMSA	−3.207383**	−2.888	0.0048	−0.199365**	−4.306	0.0001
MSA	−0.031874	−0.344	0.7319	−0.082927**	−2.076	0.0387
F-VALUE		9.38			98.441	
R-SQ		0.4578			0.747	
ADJ. R-SQ		0.409			0.7395	
N		110			310	
CON. INDEX		28.8806			12.8756	

NOTE 1: * = significance at the .10 level, ** = significance at the .05 level.
NOTE 2: The coefficients for TYGSI95, TYGL95, Y88, and MW88 are multiplied by 1,000 persons.

Table 4.10 The Determinants of the Degree of Diversity:
Cross-sectional Analysis: Small vs. Large Firms

DEP. VARIABLE: DIVERSITY (D95)

	MAs & COUNTIES			MUNICIPALITIES		
VAR.	COEFFICIENTS	T STAT.	PROB> |T|	COEFFICIENTS	T STAT.	PROB> |T|
INTERCEP	0.845777**	7.116	0.0001	0.958359**	14.263	0.0001
TYGSM95	−0.171	−0.923	0.3585	−0.35**	−3.647	0.0003
TYGL95	−0.018239	−1.383	0.1698	0.015262**	2.909	0.0039
Y88	−0.001327	−0.173	0.8634	−0.004552	−0.836	0.4036
MW88	−0.002449	−0.58	0.5629	−0.002782	−1.138	0.2562
TYGOM95	−0.056855**	−2.66	0.0091	−0.002126	−0.718	0.4735
YOM88	−0.007717	−1.254	0.2128	0.000641*	1.696	0.0909
COLL	−0.006422	−1.002	0.3189	−0.006833**	−4.231	0.0001
ACCESS	0.000185	0.818	0.4153	−0.000066602	−0.368	0.7134
COMP88	−0.001647*	−1.664	0.0994	−0.001202	−1.512	0.1315
CMSA	−1.493803	−1.028	0.3063	−0.125573**	−2.477	0.0138
MSA	−0.192006**	−2.196	0.0305	−0.062236	−1.521	0.1293
F-VALUE		4.8			9.602	
R-SQ		0.3501			0.2617	
ADJ. R-SQ		0.2772			0.2344	
N		110			310	
CON. INDEX		44.6036			13.54	

NOTE 1: * = significance at the .10 level, ** = significance at the .05 level.
NOTE 2: The coefficients for TYGSI95, TYGL95, Y88, MW88, TYGOM95, and YOM88 are multiplied by 1,000 persons.

4.2.3 The Effects of Historical Conditions on High-tech Employment Growth: Time-series Analysis

This section reports the empirical results for the simultaneous equation model for small and large firms. Current employment growth of small and large firms (YGSM95 and YGL95) is assumed to be simultaneously determined with current dynamic externalities (SP95 and D95), and current manufacturing wage (MW95). Equations (14), (15), (16) and (17) are relevant here. Equations (14) and (15) state that current high-tech employment growth of small and large firms (YGSM95 and YGL95), respectively, are a function of yearly differences of specialization (DSP), yearly differences of diversity (DD), persistence (initial level) of high-tech employment (Y88), controlling for regional variables and dummies. By comparing the coefficients of these equations, the differences by firm size can be distinguished.

Regression results are laid out in Table 4.11 and Table 4.12. Since the Hausman specification test also shows simultaneity between the fitted values from the model and the error terms, the results reported are two-stage least-squares estimates. The major findings are as follows:

1. Yearly differences of the specialization variables (DSP) had positively significant effects on current employment growth of small and large firms (YGSM95 and YGL95). Larger firms seem to have stronger MAR effects, at both levels of observation. Figure 4.4 clearly shows the lagged effects of specialization for current high-tech employment. As Henderson (1994) argues, lagged specialization variables have about 5–6 years of time-lag.
2. Unexpectedly, yearly differences of the diversity variables (DD) were positively correlated with current high-tech employment growth of small and large firms. Unlike the cross-sectional analysis, Jacobs effects are not observable. It seems that diversity had longer term effects, but the effects of yearly differences of diversity were minimal. Diversity effects might be overridden by negative urbanization effects.
3. For each group, a very high degree of persistence was observed. The coefficients of this variable were positively significant in metropolitan areas and municipalities. Large firms grew significantly in CMSAs for the period of 1988–1995.

4.2.4 The Determinants of the Degree of Dynamic Externalities: Time-series Analysis

This section deals with equations (16) and (17). In these equations, current specialization (SP95) and current diversity (D95) are assumed to depend on total high-tech employment growth of small and large firms (TYGSM95 and TYGL95, respectively) and persistence (initial level) of high-tech employment (Y88), controlling for regional variables and dummies. As noted, yearly differences of manufacturing wages (DMW) are included as control variables.

Table 4.13 shows the two-stage least-squares estimates on specialization (SP95). In metropolitan areas, employment growth of large firms (YGL95) strongly affected current specialization (SP95), but when municipalities were the units of observation, unlike the cross-sectional analysis, only employment growth of small firms was positively significant. Employment growth of small firms also contributed to concentrations of high-tech industry. Control variables have the same sign as in the cross-sectional analysis. All the results confirm the findings in the cross-sectional analysis.

Table 4.14 shows the two-stage least-squares estimates on diversity (D95). Employment growth of small firms (TYGSM95) was negatively associated with current diversity (D95). Growth of small firms led to diversified environments. Other control variables that affect diversity, such as the employment growth of manufacturing industries (TYGOM95), availability of highly-skilled labor (COLL), and the initial competition (COMP88), show the expected signs, although they are less significant than in the cross-sectional analysis.

**Table 4.11 The Effects of Historical Conditions on High-tech Employment Growth:
Time-series Analysis: Small vs. Large Firms
Two-Stage Least Squares Estimates: MAs and Counties**

DEP. VARIABLE: HIGH-TECH EMPLOYMENT GROWTH (YGSM95 and YGL95)

	MAs & COUNTIES			MUNICIPALITIES		
VAR.	COEFFICIENTS	T STAT.	PROB> |T|	COEFFICIENTS	T STAT.	PROB> |T|
INTERCEP	−53.940225	−0.734	0.4656	−1086.504606	−0.814	0.418
DSP95	690.598653*	1.937	0.0566	13903**	2.149	0.0349
DSP94	79.386931	0.641	0.5234	1555.176425	0.692	0.4909
DSP93	141.312057	1.283	0.2035	2867.962312	1.435	0.1554
DSP92	190.844265	1.467	0.1467	3804.966337	1.612	0.1112
DSP91	116.896104	1.094	0.2776	2487.980237	1.283	0.2035
DSP90	−45.202734	−0.778	0.4391	−671.254853	−0.637	0.5263
DSP89	−52.438347	−1.159	0.2502	−813.70943	−0.991	0.3248
DD95	525.233932	1.309	0.1946	9425.74308	1.295	0.1994
DD94	283.341763*	1.72	0.0896	5039.384996*	1.686	0.0959
DD93	116.985155	1.088	0.2802	2114.154122	1.084	0.2821
DD92	131.83983	0.986	0.3272	2003.701303	0.826	0.4114
DD91	194.278788*	1.67	0.0991	3101.996987	1.47	0.1458
DD90	166.689497*	1.879	0.0642	2559.550715	1.59	0.1161
DD89	66.321918	0.825	0.4122	1082.950292	0.742	0.4603
Y88	0.01154**	7.724	0.0001	0.11696**	4.315	0.0001
MW95	0.002494	1.272	0.2074	0.045838	1.288	0.2016
COLL	6.619431*	1.724	0.0888	124.932996*	1.794	0.0769
ACCESS	−0.089249	−0.552	0.5823	−1.625755	−0.555	0.5808
COMP88	0.360201	0.392	0.6965	4.503963	0.27	0.788
CMSA	−191.481311	−0.967	0.3369	8614.26642**	2.397	0.0191
MSA	7.978912	0.188	0.851	−5.17313	−0.007	0.9946
F-VALUE		14.172			13.818	
R-SQ		0.8009			0.7968	
ADJ. R-SQ		0.7444			0.7391	
N		96			96	

NOTE: * = significance at the .10 level, ** = significance at the .05 level.

**Table 4.12 The Effects of Historical Conditions on High-tech Employment Growth:
Time-series Analysis: Small vs. Large Firms
Two-Stage Least Squares Estimates: Municipalities**

DEP. VARIABLE: HIGH-TECH EMPLOYMENT GROWTH (YGSM95 and YGL95)

VAR.	SMALL COEFFICIENTS	T STAT.	PROB> |T|	LARGE COEFFICIENTS	T STAT.	PROB> |T|
INTERCEP	−12.493579	−0.59	0.5559	530.147892	1.151	0.2508
DSP95	370.875684**	2.153	0.0322	14950**	3.991	0.0001
DSP94	−75.69159**	−2.193	0.0292	1384.823021*	1.845	0.0661
DSP93	−0.490793	−0.02	0.984	1431.067145**	2.684	0.0077
DSP92	52.554436	1.489	0.1376	2262.22727**	2.948	0.0035
DSP91	42.858541	0.807	0.4202	4802.528933**	4.16	0.0001
DSP90	28.030634	0.741	0.4596	3086.442135**	3.75	0.0002
DSP89	−48.900926**	−2.725	0.0069	911.701281**	2.336	0.0202
DD95	257.405984	1.486	0.1384	1155.622013	0.307	0.7592
DD94	137.027923*	1.668	0.0966	1334.31427	0.747	0.4558
DD93	125.606553**	2.074	0.039	1527.899215	1.16	0.2469
DD92	88.305495*	1.965	0.0505	480.257829	0.491	0.6235
DD91	60.029315	1.515	0.1309	65.775863	0.076	0.9392
DD90	85.922556**	2.265	0.0243	721.966407	0.875	0.3822
DD89	35.112505	1.274	0.2036	1114.236029*	1.86	0.064
Y88	0.009543**	8.937	0.0001	0.30616**	13.187	0.0001
MW95	0.000871	1.409	0.1601	−0.005694	−0.424	0.6722
COLL	0.632214	1.301	0.1944	6.730551	0.637	0.5247
ACCESS	0.032639	0.565	0.5726	−0.645736	−0.514	0.6077
COMP88	0.025089	0.099	0.921	−3.255801	−0.593	0.5539
CMSA	−1.666189	−0.124	0.9016	−417.720765	−1.426	0.1549
MSA	3.63822	0.307	0.7591	32.470855	0.126	0.8998
F-VALUE		12.577			23.593	
R-SQ		0.4917			0.6447	
ADJ. R-SQ		0.4526			0.6174	
N		295			295	

NOTE: * = significance at the .10 level, ** = significance at the .05 level.

[MAs & COUNTIES]

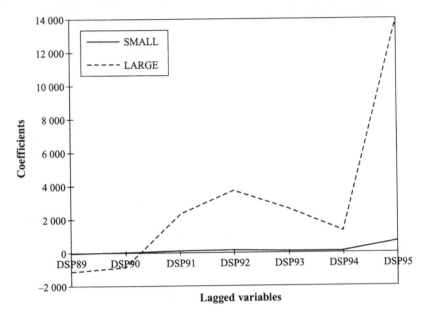

[MUNICIPALITIES]

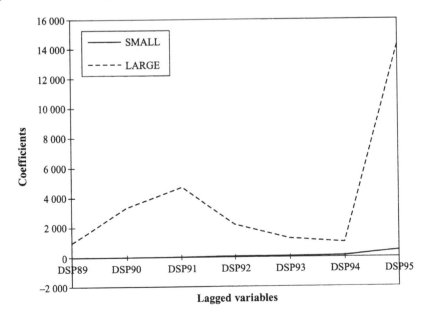

**Figure 4.4 The Lagged Effects of Specialization on High-tech Employment
Growth: Small vs. Large Firms**

Table 4.13 The Determinants of the Degree of Specialization:
Time-series Analysis: Small vs. Large Firms
Two-Stage Least Squares Estimates

DEP. VARIABLE: SPECIALIZATION (SP95)

	MAs & COUNTIES			MUNICIPALITIES		
VAR.	COEFFICIENTS	T STAT.	PROB> |T|	COEFFICIENTS	T STAT.	PROB> |T|
INTERCEP	0.027592	0.162	0.8719	0.117876	1.382	0.168
TYGSM95	−0.915	−1.372	0.174	1.193**	3.146	0.0018
TYGL95	0.069173**	2.801	0.0064	−0.000184	−0.013	0.9896
Y88	0.042934*	1.808	0.0743	−0.013381	−0.839	0.4021
DMW95	0.002013	0.162	0.872	−0.00184	−0.297	0.7669
DMW94	−0.015217	−0.665	0.5082	−0.027797*	−1.788	0.0748
DMW93	−0.02042	−0.835	0.4063	−0.017229	−0.923	0.3571
DMW92	−0.007338	−0.402	0.6888	−0.007185	−0.562	0.5743
DMW91	0.006148	0.31	0.757	−0.006184	−0.508	0.6118
DMW90	0.024621	1.021	0.3104	0.013093	0.786	0.4327
DMW89	0.0266	1.018	0.3118	0.010549	0.732	0.4646
COLL	0.011977	1.181	0.2411	0.002825	1.215	0.2252
ACCESS	0.000746*	1.932	0.0569	0.000399	1.449	0.1486
COMP88	−0.00082	−0.489	0.626	−0.00039	−0.351	0.7258
CMSA	−0.76369	−0.3	0.7652	−0.182134**	−2.929	0.0037
MSA	0.05835	0.493	0.6232	−0.073177	−1.303	0.1935
F-VALUE		4.317			28.622	
R-SQ		0.4473			0.6061	
ADJ. R-SQ		0.3437			0.5849	
N		96			295	

NOTE 1: * = significance at the .10 level, ** = significance at the .05 level.
NOTE 2: The coefficients for TYG95, Y88, and DMW are multiplied by 1,000 persons.

Table 4.14 The Determinants of the Degree of Diversity:
Time-series Analysis: Small vs. Large Firms
Two-Stage Least Squares Estimates

DEP. VARIABLE: DIVERSITY (D95)

	MAs & COUNTIES			MUNICIPALITIES		
VAR.	COEFFICIENTS	T STAT.	PROB> |T|	COEFFICIENTS	T STAT.	PROB> |T|
INTERCEP	0.761582**	5.757	0.0001	0.836337**	12.647	0.0001
TYGSM95	−0.962	−1.413	0.1616	−0.528*	−1.831	0.0682
TYGL95	0.018388	0.49	0.6256	0.010986	0.869	0.3854
Y88	0.026122	1.073	0.2864	0.00587	0.457	0.648
DMW95	−0.009353	−0.984	0.3281	0.001978	0.406	0.685
DMW94	0.022994	1.325	0.1892	0.018548	1.491	0.137

DMW93	0.024206	1.266	0.2094	0.019157	1.356	0.1762
DMW92	−0.010211	−0.734	0.4649	−0.006793	−0.716	0.4744
DMW91	−0.014908	−0.983	0.3285	0.005443	0.548	0.5841
DMW90	−0.016289	−0.885	0.3789	0.006158	0.441	0.6593
DMW89	−0.005889	−0.284	0.7772	−0.002934	−0.275	0.7836
TYGOM95	−0.037837	−0.862	0.3912	−0.007911	−1.303	0.1938
YOM88	0.001203	0.119	0.9059	0.000453	1.105	0.2701
COLL	0.000049912	0.006	0.9954	−0.006152**	−3.569	0.0004
ACCESS	−0.000035819	−0.116	0.9078	0.000027061	0.128	0.8979
COMP88	−0.001547	−1.136	0.2595	−0.001313	−1.564	0.119
CMSA	0.819036	0.329	0.743	−0.122524**	−2.301	0.0221
MSA	−0.184117*	−1.876	0.0644	−0.049151	−1.107	0.2695
F-VALUE		2.131			5.008	
R-SQ		0.3171			0.2351	
ADJ. R-SQ		0.1683			0.1882	
N		96			295	

NOTE 1: * = significance at the .10 level, ** = significance at the .05 level.
NOTE 2: The coefficients for TYG95, Y88, DMW, TYGOM95, and YOM88 are multiplied by 1,000 persons.

In summary, hypothesis 2 – that the growth of small firms is more likely to be driven by endogenous processes than the growth of large firms – is true for the Jacobs effects. Small firms have larger Jacobs effects than large firms. On the other hand, this hypothesis is not true for MAR effects. Large firms have larger MAR effects than small firms. These findings match recent trends. As the Economist (Feb. 18, 1995: 63) argues, economies of scale can be realized by ever-smaller businesses due to flexible manufacturing techniques and the tumbling cost of computing. Small firms tend to be good innovators and spend relatively large sums on R&D. However, since they lack an internal knowledge network, more diversified environments are likely to be necessary for small firms to grow. High-tech enclaves can help small firms achieve critical mass in everything.

4.3 The Differences in Regional High-tech Employment Effects of Dynamic Externalities by Organizational Type

I now report on the cross-sectional and simultaneous equation tests of the hypothesis that the growth of single-location firms is more likely to be driven by endogenous processes than is the growth of multi-locational firms. First, I discuss in sections 4.3.1 and 4.3.2 the solutions to equations (18), (19), (20) and (21). Then I discuss the solutions to equations (22), (23), (24) and (25) in sections 4.3.3 and 4.3.4. In these tests, basically, I use the same model that is used in the previous section.

4.3.1 The Effects of Historical Conditions on High-tech Employment Growth: Cross-sectional Analysis

Equations (18) and (19) address the effects of historical conditions on employment growth of single and multi-locational establishments. In the equations, high-tech employment growth of single and multi-locational establishments (TYGSM95 and TYGL95, respectively) is a function of past specialization (SP88), past diversity (D88), and persistence (initial level) of high-tech employment (Y88), controlling for regional variables and dummies. By comparing the coefficients of the explanatory variables, I explore the differences in the effects of historical conditions on high-tech employment growth.

The regression results are laid out in Tables 4.15 and 4.16. Table 4.15 reports the coefficients when metropolitan areas and non-metropolitan counties are the units of observation. Table 4.16 reports the coefficients for municipalities as the units of observation. There are several important findings:

1. The degree of past specialization (SP88) strongly affects both types of employment growth (TYGSI95 and TYGM95) in Texan metropolitan areas. The coefficients for these variables are positively significant and quantitatively large. MAR effects are a little greater for multi-locational establishments. However, when municipalities are used as the units of observation, the effects of specialization are not readily observable.

Table 4.15 The Effects of Historical Conditions on High-tech Employment Growth: Cross-sectional Analysis: Single vs. Multi-locational Establishments MAs and Counties

DEP. VARIABLE: HIGH-TECH EMPLOYMENT GROWTH (TYGSI95 and TYGM95)

VAR.	SINGLE			MULTI-LOCATIONAL		
	COEFFICIENTS	T STAT.	PROB> |T|	COEFFICIENTS	T STAT.	PROB> |T|
INTERCEP	−2664.988483*	−1.834	0.0696	−2208.653816	−1.564	0.121
SP88	3033.998376**	5.053	0.0001	2344.856782**	4.017	0.0001
D88	157.79935	0.132	0.8955	91.215593	0.078	0.9378
Y88	−0.121085**	−4.574	0.0001	−0.083798**	−3.256	0.0015
MW88	0.034929	0.957	0.341	0.072313**	2.037	0.0443
COLL	115.166758*	2.002	0.048	0.051571	0.001	0.9993
ACCESS	0.624702	0.308	0.7589	3.041246	1.541	0.1265
COMP88	0.482034	0.053	0.9578	5.199607	0.589	0.5574
CMSA	63033**	19.856	0.0001	60688**	19.663	0.0001
MSA	88.950449	0.117	0.9072	1889.175475**	2.554	0.0122
F-VALUE		149.185			152.92	
R-SQ		0.9307			0.9323	
ADJ. R-SQ		0.9244			0.9262	
N		110			110	
CON. INDEX		19.0159			19.0159	

NOTE: * = significance at the .10 level, significance at the .05 level.

Table 4.16 The Effects of Historical Conditions on High-tech Employment Growth:
Cross-sectional Analysis: Single vs. Multi-locational Establishments
Municipalities

DEP. VARIABLE: HIGH-TECH EMPLOYMENT GROWTH (TYGSI95 and TYGM95)

	SINGLE			MULTI-LOCATIONAL		
VAR.	COEFFICIENTS	T STAT.	PROB> |T|	COEFFICIENTS	T STAT.	PROB> |T|
INTERCEP	24.136665	0.023	0.9813	−400.693856	−0.296	0.7676
SP88	−230.169885	−0.481	0.6311	784.815093	1.247	0.2132
D88	−56.634743	−0.078	0.9378	−685.061595	−0.719	0.4728
Y88	0.606364**	12.536	0.0001	0.380277**	5.984	0.0001
MW88	0.007616	0.308	0.7582	0.032395	0.998	0.3193
COLL	−3.87733	−0.232	0.8163	3.873006	0.177	0.8598
ACCESS	−0.318352	−0.172	0.8633	2.202896	0.907	0.3649
COMP88	2.085469	0.256	0.7979	−0.400735	−0.037	0.9701
CMSA	−193.236461	−0.391	0.696	41.346118	0.064	0.9493
MSA	−67.789482	−0.162	0.8713	15.913451	0.029	0.9769
F-VALUE		46.513			17.097	
R-SQ		0.5825			0.339	
ADJ. R-SQ		0.57			0.3192	
N		310			310	
CON. INDEX		18.8515			18.8515	

NOTE: * = significance at the .10 level, ** = significance at the .05 level.

2. The degree of past diversity (D88) did not have any significant effects on either type of employment growth. Jacobs effects was not observed for either organizational type.
3. Initial high-tech level (Y88) had negatively significant effects on both types of employment growth in metropolitan areas, but when municipalities are used as the units of observations, this variable had the opposite effect. This implies that single and multi-locational firms both as a whole grew slower in heavily concentrated metropolitan areas.
4. Manufacturing wage (MW88) was positively significant only for multi-locational establishments, whereas availability of highly-skilled labor (COLL) was positively significant only for single firms. Considering that wage premiums could reflect technical expertise, both organizational types are thought to react positively to human capital.

4.3.2 The Determinants of the Degree of Dynamic Externalities:
Cross-sectional Analysis

The next question deals with the differences in the determinants of the degree of dynamic externalities by organizational type. I focus on the effects of high-tech employment growth of single and multi-locational establishments on the degree of dynamic externalities. Equations (20) and (21) are relevant here. In these equations, current dynamic externalities (either specialization or diversity) are a function of employment growth of single and multi-locational establishments (TYGSI95 and TYGM95, respectively), and persistence (initial level) of high-tech employment (Y88), controlling for regional variables and dummies.

The regression results on current specialization are shown in Table 4.17. Current specialization was affected mainly by employment growth of multi-locational establishments. The coefficients for employment growth of multi-locational establishments (TYGM95) were positively significant in metropolitan areas and municipalities. In contrast, the effects of employment growth of single establishments (TYGSI95) was insignificant. These findings mean that multi-locational establishments were the major forces affecting current specialization for the period of 1988 to 1995 in Texan high-tech industry. Other control variables shows the same results as in total products or as in small vs. large establishments.

Table 4.18 also shows the regression results on current diversity (D95). Neither employment growth of single establishments (TYGSM95) nor employment growth of multi-locational establishments (TYGM95) is responsible for current diversity (D95).

4.3.3 The Effects of Historical Conditions on High-tech Employment Growth: Time-series Analysis

I now turn to the simultaneous equation model for single and multi-locational establishments. Current employment growth of single and large establishments (YGSI95 and YGM95, respectively) is assumed to be simultaneously determined with current dynamic externalities (SP95 and D95), and current manufacturing wage (MW95). Equations (22), (23), (23) and (24) are relevant here. Equations (22) and (23) state that current high-tech employment growth of single establishments and multi-locational establishments (YGSI95 and YGM95), respectively, are a function of yearly differences of specialization (DSP), yearly differences of diversity (DD), persistence (initial level) of high-tech employment (Y88), controlling for regional variables and dummies. By comparing the coefficients of these equations, the differences by organizational type can be distinguished.

Table 4.17 The Determinants of the Degree of Specialization: Cross-sectional Analysis: Single and Multi-locational Establishments

| | DEP. VARIABLE: SPECIALIZATION (SP95) | | | | | |
| | MAs & COUNTIES | | | MUNICIPALITIES | | |
VAR.	COEFFICIENTS	T STAT.	PROB> |T|	COEFFICIENTS	T STAT.	PROB> |T|
INTERCEP	−0.260463*	−1.738	0.0853	−0.046506	−0.722	0.4706
TYGSI95	0.017367	0.827	0.41	0.003299	0.539	0.5901
TYGM95	0.041037*	1.818	0.0721	0.056106**	12.101	0.0001
Y88	0.008703**	2.42	0.0173	0.044101**	10.354	0.0001
MW88	0.016098**	3.039	0.003	0.006676**	2.859	0.0045

COLL	0.010043	1.197	0.2341	0.004047**	2.635	0.0088
ACCESS	0.00073*	2.485	0.0146	0.000446**	2.566	0.0108
COMP88	−0.001158	−0.915	0.3624	−0.000336	−0.442	0.6591
CMSA	−3.324433**	−3.704	0.0003	−0.217681**	−4.83	0.0001
MSA	−0.056251	−0.569	0.5708	−0.096205	−2.471	0.014
F-VALUE		9.444			104.959	
R-SQ		0.4595			0.759	
ADJ. R-SQ		0.4108			0.7517	
N		110			310	
CON. INDEX		17.701			12.7127	

NOTE 1: * = significance at the .10 level, ** = significance at the .05 level.
NOTE 2: The coefficients for TYGSI95, TYGL95, Y88, and MW88 are multiplied by 1,000
persons.

Table 4.18 The Determinants of the Degree of Diversity:
         Cross-sectional Analysis: Single and Multi-locational
         Establishments

DEP. VARIABLE: DIVERSITY (D95)

	MAs & COUNTIES			MUNICIPALITIES		
VAR.	COEFFICIENTS	T STAT.	PROB> |T|	COEFFICIENTS	T STAT.	PROB> |T|
INTERCEP	0.850542**	7.132	0.0001	0.951364**	13.842	0.0001
TYGSI95	−0.022729	−1.195	0.2349	−0.000765	−0.11	0.9122
TYGM95	−0.029576	−1.165	0.2468	0.004057	0.721	0.4715
Y88	−0.006807	−1.639	0.1044	−0.0166**	−3.643	0.0003
MW88	−0.00237	−0.559	0.5773	−0.002331	−0.933	0.3513
TYGOM95	−0.06222**	−2.785	0.0064	−0.002436	−0.805	0.4215
YOM88	−0.008785	−1.355	0.1787	0.000576	1.495	0.1361
COLL	−0.007293	−1.113	0.2684	−0.00742**	−4.512	0.0001
ACCESS	0.000207	0.906	0.3674	−0.000053891	−0.291	0.7715
COMP88	−0.001648	−1.659	0.1004	−0.001129	−1.39	0.1656
CMSA	−2.047179	−1.322	0.1893	−0.135575**	−2.62	0.0092
MSA	−0.189883**	−2.156	0.0335	−0.069585**	−1.665	0.0969
F-VALUE		4.718			8.042	
R-SQ		0.3462			0.2289	
ADJ. R-SQ		0.2729			0.2004	
N		110			310	
CON. INDEX		42.5621			13.3465	

NOTE 1: * = significance at the .10 level, ** = significance at the .05 level.
NOTE 2: The coefficients for TYGSI95, TYGL95, Y88, MW88, TYGOM95, and YOM88 are
        multiplied by 1,000 persons.

Regression results are laid out in Table 4.19 and Table 4.20. The Hausman specification test shows simultaneity between the fitted values from the model and the error terms, and the results reported therefore are two-stage least-squares estimates. The major findings are as follows:

1. Yearly differences of the specialization variables (DSP) had positively significant effects on current employment growth of single and multi-locational establishments (YGSI95 and YGM95). In metropolitan areas, MAR effects are almost the same for both organizational types. However, when municipalities are used as the units of observation, multi-locational establishments have more stronger MAR effects, and the time-lags seem to be a little longer. Figure 4.5 clearly shows the lagged effects of specialization on current high-tech employment. This finding implies that multi-locational establishments adjust more quickly to specialized environments.

Table 4.19 The Effects of Historical Conditions on High-tech Employment Growth: Time-series Analysis: Single vs. Multi-locational Establishments Two-Stage Least Squares Estimates: MAs and Counties

DEP. VARIABLE: HIGH-TECH EMPLOYMENT GROWTH (YGSI95 and YGM95)

	MAs & COUNTIES			MUNICIPALITIES		
VAR.	COEFFICIENTS	T STAT.	PROB> |T|	COEFFICIENTS	T STAT.	PROB> |T|
INTERCEP	−1438.33538	−1.645	0.1043	297.89055	0.291	0.7715
DSP95	7472.04556*	1.762	0.0821	7121.878067	1.437	0.1549
DSP94	578.073449	0.393	0.6957	1056.489907	0.614	0.5412
DSP93	1395.279921	1.065	0.2902	1613.994448	1.054	0.2952
DSP92	2353.105937	1.521	0.1326	1642.704665	0.908	0.3667
DSP91	959.423763	0.755	0.4527	1645.452577	1.108	0.2716
DSP90	−140.816838	−0.204	0.8391	−575.640749	−0.713	0.4783
DSP89	−610.742682	−1.135	0.26	−255.405095	−0.406	0.6859
DD95	−1077.925525	−0.226	0.8219	11029*	1.977	0.0517
DD94	1299.42493	0.663	0.5091	4023.301829*	1.757	0.083
DD93	529.59766	0.414	0.68	1701.541617	1.138	0.2587
DD92	−15.434416	−0.01	0.9923	2150.975549	1.158	0.2508
DD91	1599.544252	1.156	0.2513	1696.731523	1.049	0.2974
DD90	1167.150405	1.106	0.2722	1559.089807	1.264	0.2101
DD89	778.848413	0.814	0.418	370.423797	0.331	0.7413
Y88	0.085829**	4.831	0.0001	0.042671**	2.055	0.0435
MW95	0.035578	1.526	0.1314	0.012755	0.468	0.6412
COLL	120.61672**	2.642	0.01	10.935707	0.205	0.8382
ACCESS	−2.108563	−1.098	0.276	0.393559	0.175	0.8614
COMP88	−10.261017	−0.938	0.3513	15.125182	1.183	0.2406
CMSA	−2088.31655	−0.886	0.3783	10511**	3.817	0.0003
MSA	−810.438108	−1.61	0.1117	813.24389	1.382	0.1711

F-VALUE	7.068	9.188
R-SQ	0.6673	0.7228
ADJ. R-SQ	0.5729	0.6441
N	96	96

NOTE: * = significance at the .10 level, ** = significance at the .05 level.

Table 4.20 The Effects of Historical Conditions on High-tech Employment Growth: Time-series Analysis: Single vs. Multi-locational Establishments Two-Stage Least Squares Estimates: Municipalities

DEP. VARIABLE: HIGH-TECH EMPLOYMENT GROWTH (YGSI95 and YGM95)

	MAs & COUNTIES			MUNICIPALITIES		
VAR.	COEFFICIENTS	T STAT.	PROB> |T|	COEFFICIENTS	T STAT.	PROB> |T|
INTERCEP	251.502634	0.782	0.435	266.151679	0.985	0.3257
DSP95	6939.162874**	2.653	0.0085	8381.779483**	3.813	0.0002
DSP94	403.659407	0.77	0.4419	905.472024**	2.056	0.0408
DSP93	683.222427*	1.835	0.0676	747.353925**	2.388	0.0176
DSP92	1255.861754**	2.344	0.0198	1058.919952**	2.352	0.0194
DSP91	1935.932464**	2.401	0.017	2909.45501**	4.295	0.0001
DSP90	1044.975603*	1.818	0.0702	2069.497165**	4.284	0.0001
DSP89	−113.926496	−0.418	0.6762	976.726851**	4.265	0.0001
DD95	1269.952866	0.483	0.6296	143.075131	0.065	0.9484
DD94	1045.94397	0.838	0.4026	425.398223	0.406	0.6853
DD93	1138.367658	1.238	0.2168	515.13811	0.667	0.5056
DD92	533.311236	0.781	0.4353	35.252088	0.061	0.951
DD91	162.414888	0.27	0.7874	−36.60971	−0.072	0.9423
DD90	745.216972	1.294	0.1968	62.671991	0.129	0.8971
DD89	699.705528*	1.672	0.0956	449.643006	1.279	0.202
Y88	0.153618**	9.475	0.0001	0.162086**	11.896	0.0001
MW95	0.003791	0.404	0.6866	−0.008614	−1.092	0.2758
COLL	3.559568	0.482	0.6299	3.803197	0.613	0.5402
ACCESS	−0.887606	−1.012	0.3126	0.274509	0.372	0.7099
COMP88	−0.319148	−0.083	0.9338	−2.911564	−0.903	0.3673
CMSA	−378.237784*	−1.85	0.0655	−41.149171	−0.239	0.811
MSA	−69.52215	−0.386	0.6996	105.631226	0.698	0.4856
F-VALUE		12.798			19.283	
R-SQ		0.4961			0.5973	
ADJ. R-SQ		0.4573			0.5663	
N		295			295	

NOTE: * = significance at the .10 level, ** = significance at the .05 level.

2. However, as in the cross-sectional analysis, yearly differences of the diversity variables (DD) did not affect current employment growth of single and multi-locational establishments. Jacobs effects are not observable.
3. At both spatial scales, strong persistence effects were observed. Initial high-tech employment levels positively affected both types of employment growth. As the regional dummies indicate, multi-locational establishments grew rapidly in CMSAs, but single firms grew at a slower rate.
4. Availability of highly-skilled labor was positively significant for single firms in metropolitan areas. For single firms, the human capital factor was more important than in multi-locational establishments.

4.3.4 The Determinants of the Degree of Dynamic Externalities: Time-series Analysis

I now deal with equations (23) and (24). In these equations, current specialization (SP95) and current diversity (D95) are assumed to depend on total high-tech employment growth of single and multi-locational establishments (TYGSI95 and TYGM95, respectively) and persistence (initial level) of high-tech employment (Y88), controlling for regional variables and dummies. As noted, yearly differences of manufacturing wages (DMW) also are included as control variables.

Table 4.21 shows the two-stage least-squares estimates on specialization (SP95). In metropolitan areas, employment growth of multi-locational establishments (YGM95) strongly affected current specialization (SP95), but when municipalities were the units of observation, only employment growth of single firms was positively significant. Both types of employment growth contributed to concentrations of high-tech industry.

Table 4.22 shows the two-stage least-squares estimates on diversity (D95). In metropolitan areas, only employment growth of multi-locational establishments (TYGM95) was negatively associated with current diversity (D95). When municipalities are used as the unit of observations, employment growth of single firms was almost significantly and negatively associated with current diversity (D95).

In summary, that part of hypothesis 2 that stated that the growth of single firms is more likely to be driven by endogenous processes than is the growth of multi-locational establishments, is not true. Rather, multi-locational establishments displays somewhat stronger MAR effects than single firms. This implies that multi-locational establishments responded more quickly to dynamic externalities.

[MAs & COUNTIES]

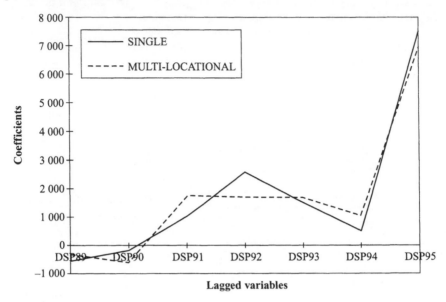

[MUNICIPALITIES]

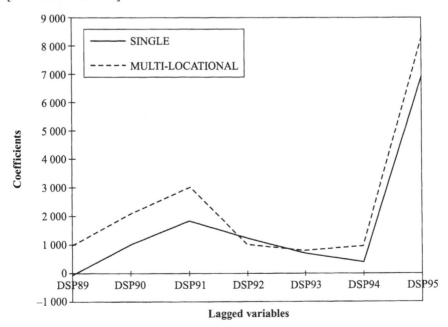

Figure 4.5 The Lagged Effects of Specialization on High-tech Employment Growth: Single and Multi-locational Establishments

**Table 4.21 The Determinants of the Degree of Specialization:
Time-series Analysis: Single vs. Multi-locational Establishments
Two-Stage Least Squares Estimates**

DEP. VARIABLE: SPECIALIZATION (SP95)

VAR.	MAs & COUNTIES			MUNICIPALITIES		
	COEFFICIENTS	T STAT.	PROB> |T|	COEFFICIENTS	T STAT.	PROB> |T|
INTERCEP	0.005042	0.031	0.9753	0.093418	1.156	0.2486
TYGSI95	−0.031571	−0.834	0.4069	0.095167**	4.759	0.0001
TYGM95	0.104**	2.765	0.0071	0.008116	0.674	0.5009
Y88	0.007866*	1.951	0.0546	0.012399	1.381	0.1683
DMW95	0.000597	0.051	0.9595	−0.001385	−0.236	0.8133
DMW94	−0.019522	−0.903	0.3692	−0.031097**	−2.117	0.0351
DMW93	−0.024266	−1.046	0.2986	−0.02289	−1.31	0.1912
DMW92	−0.010456	−0.604	0.5476	−0.003132	−0.26	0.7951
DMW91	−0.002569	−0.135	0.8933	−0.003104	−0.271	0.7862
DMW90	0.026441	1.174	0.244	0.012531	0.797	0.4261
DMW89	0.034271	1.44	0.1537	0.005651	0.418	0.676
COLL	0.016525	1.6	0.1136	0.005187**	2.435	0.0155
ACCESS	0.000577	1.509	0.1352	0.000457*	1.749	0.0814
COMP88	−0.001514	−0.94	0.3498	−0.000693	−0.661	0.5094
CMSA	3.879551**	−3.507	0.0007	−0.137107**	−2.374	0.0183
MSA	−0.147271	−1.225	0.2244	−0.040799	−0.775	0.4388
F-VALUE		4.932			32.013	
R-SQ		0.4805			0.6325	
ADJ. R-SQ		0.383			0.6127	
N		96			295	

NOTE 1: * = significance at the .10 level, ** = significance at the .05 level.
NOTE 2: The coefficients for TYGSI95, TYGL95, Y88, and MW88 are multiplied by 1,000
persons.

**Table 4.22 The Determinants of the Degree of Diversity:
Time-series Analysis: Single vs. Multi-locational Establishments
Two-Stage Least Squares Estimates**

DEP. VARIABLE: DIVERSITY (D95)

VAR.	MAs & COUNTIES			MUNICIPALITIES		
	COEFFICIENTS	T STAT.	PROB> |T|	COEFFICIENTS	T STAT.	PROB> |T|
INTERCEP	0.809101	5.321	0.0001	0.835577**	11.915	0.0001
TYGSI95	0.09411	1.388	0.1691	−0.041126	−1.544	0.1237
TYGM95	−0.204**	−2.051	0.0436	0.003591	0.317	0.7515
Y88	−0.018743**	−1.995	0.0495	−0.002901	−0.288	0.7735
DMW95	−0.007179	−0.646	0.5205	0.002831	0.544	0.587

DMW94	0.026773	1.324	0.1895	0.023518*	1.719	0.0867
DMW93	0.040352*	1.765	0.0815	0.019892	1.331	0.1842
DMW92	−0.000309	−0.019	0.9852	−0.009376	−0.927	0.3549
DMW91	−0.009511	−0.532	0.5966	0.006791	0.632	0.5281
DMW90	−0.008175	−0.389	0.6985	0.009824	0.643	0.5205
DMW89	0.007629	0.344	0.7315	−0.001203	−0.107	0.9148
TYGOM95	−0.127**	−2.042	0.0445	−0.011659	−1.574	0.1166
YOM88	0.011201	0.827	0.411	0.000283	0.642	0.5216
COLL	−0.015001	−1.428	0.1572	−0.007276**	−4.053	0.0001
ACCESS	0.000497	1.219	0.2264	0.000029296	0.131	0.8956
COMP88	−0.00161	−1.021	0.3102	−0.001261	−1.421	0.1563
CMSA	−7.666826**	−2.068	0.0419	−0.128838**	−2.3	0.0222
MSA	−0.129627	−1.097	0.2762	−0.054348	−1.164	0.2454
F-VALUE		1.699			4.424	
R-SQ		0.2702			0.2135	
ADJ. R-SQ		0.1112			0.1653	
N		96			295	

NOTE 1: * = significance at the .10 level, ** = significance at the .05 level.
NOTE 2: The coefficients for TYGSI95, TYGL95, Y88, MW88, TYGOM95, and YOM88 are multiplied by 1,000 persons.

This finding seems to be inconsistent with previous studies. Harris (1986: 169) argues that agglomeration economies are less important to new branch establishments than to new independent firms. The reasons are that in many large, vertically integrated and conglomerate firms, intrafirm sourcing to geographically dispersed subsidiaries or branch plants is feasible and cost-effective, and in many high-tech industries shipping costs are relatively low. However, multi-locational establishments tend to be international and in many cases, within-firm resources could be limited.

4.4 The Differences in Regional High-tech Employment Effects of Dynamic Externalities by Product

The regression results regarding the cross-sectional and simultaneous equation models of individual high-tech product now should be discussed. These regressions test hypothesis 1 that regional growth in high-tech employment is determined by endogenous growth processes, and hypothesis 3 that endogenous growth is more likely to occur in competitive local economies rather than in local economies dominated by one or relatively few large firms. First, I discuss the solutions to equations (26), (27), and (28) in sections 4.4.1 and 4.4.2. Then, I discuss the solutions to equations (29), (30), and (31) in sections 4.4.3 and 4.4.4. These are related to testing hypothesis (1). I next focus on the effects of local competition on high-tech employment growth in section 4.4.5. The same model is adopted for each individual high-tech product. If the regional high-tech employment effects of dynamic externalities are robust, then the pattern should be consistent across high-tech products.

As noted, in this study, high-tech products in Texas are classified into 13 groups. To compare with theoretical expectations, in terms of SIC codes, these are broadly grouped into the three kinds of high-tech activities prevailed in the USA in the 1980s: *innovation-based industries* with high proportions of high-wage, non-production occupations; *intellectual-capital industries* employing high proportions of highly-skilled, technical labor; and *variety-based industries* with a high degree of product diversification. According to Pollard (1996: 8), as a whole, intellectual-capital industries and innovation-based industries seemed to be increasingly important, whereas variety-based manufacturing has seen its share of employment decline steadily over time.

4.4.1 The Effects of Historical Conditions on High-tech Employment Growth: Cross-sectional Analysis

The model to be tested is basically the same as that for total products, although the operational definitions of the variables are adjusted. Equation (26) states that high-tech employment growth (TYG95) is a function of past specialization (SP88), past diversity (D88), persistence (initial level) of high-tech employment (Y88), controlling for regional variables and dummies.

The regression results are laid out in Table 4.23. Each column reports the coefficients for high-tech product. The important findings are as follows:

1. The positive effect of past specialization (SP88) was observed only for aerospace/aircraft. In fact, a negative effect of past specialization was observed in most industries. For some products, such as chemicals, other manufacturing industries, and research, development and testing service industries, the coefficients for this variable were negatively significant. Variety-based industries tend to have strong negative effects.
2. On the other hand, the diversity effects prevail in many industries, including aerospace/aircraft, measurement and testing equipment, software, and other manufacturing industries. Intellectual-capital industries tend to have strong diversity effects.
3. The effects of the initial level of high-tech employment (Y88) were strongly significant, as expected. This variable was positively or negatively associated with employment growth of individual products (TYG95), each displaying its own growth pattern.
4. The availability of highly-skilled labor (COLL) was significant, especially in intellectual-capital industries such as measurement and testing equipment, software, and other high-tech manufacturing. This is consistent with national patterns, as Pollard (1996) argues.

Overall, the evidence is that both MAR effects and Jacobs effects were present in Texan high-tech development for the period of 1988 to 1995. The specialization and diversity patterns imply that in most industries, small firms have become more prominent.

4.4.2 The Determinants of the Degree of Dynamic Externalities: Cross-sectional Analysis

Equations (27) and (28) are relevant to the determinants of the degree of dynamic externalities. In these equations, either current specialization (SP95) or current diversity (D95) is a function of high-tech employment growth (TYG95), persistence (initial level) of high-tech employment (Y88), controlling for regional variables and dummies. The regression results are shown in Tables 4.24 and Table 4.25, respectively. The conclusions are as follows:

1. The initial employment level (Y88) and total employment growth (TYG95) of individual product did not significantly affect current specialization (SP95) and current diversity (D95), except for a few industry such as aerospace/aircraft and microelectronics industries. This finding implies that high-tech products were highly inter-linked; transformation into other products was readily achieved in adaptation to rapid economic and technical change.

Table 4.23 The Effects of Historical Conditions on High-tech Employment Growth: Cross-sectional Analysis: by Product

DEP. VARIABLE: HIGH-TECH EMPLOYMENT GROWTH (TYG95)

VAR.	INNOVATION-BASED INDUSTRIES						
	Microelectronics		Computer and Peripherals	Medical Equipment	Aerospace/Aircraft		Military Equipment
	a	b	b	b	a	b	b
INTERCEP	−2368.499	−1118.928	−2292.584	962.169	322.809	3585.788	3611.305
	(−0.509)	(−0.617)	(−0.432)	(1.387)	(0.509)	(1.284)	(1.128)
SP88	−23.727	−44.415	−304.764	−11.366	103.729*	−164.227	−291.727
	(−0.043)	(−0.268)	(−0.441)	(−1.203)	(1.92)	(−0.533)	(−1.267)
D88	−3100.141	−1532.515	−3029.890	−401.548	−165.969	−3654.337**	−2063.573
	(−1.095)	(−1.571)	(−1.281)	(−1.074)	(−0.373)	(−2.261)	(−1.207)
Y88	−0.061	−0.198**	−0.390**	1.152*	−0.306**	−0.663**	−0.405**
	(−0.439)	(−2.117)	(−2.015)	(1.91)	(−24.74)	(−6.613)	(−2.248)
MW88	0.068	0.056	0.249	−0.017	0.000	0.017	−0.045
	(0.495)	(1.047)	(1.605)	(−0.907)	(0.01)	(0.185)	(−0.48)
COLL	197.348	40.211	37.412	4.585	−5.840	8.064	11.110
	(1.46)	(1.27)	(0.571)	(0.374)	(−0.307)	(0.14)	(0.204)
ACCESS	10.981*	6.709*	−1.255	−0.692	−0.603	−2.312	−3.566
	(1.843)	(1.952)	(−0.131)	(−0.587)	(−0.562)	(−0.334)	(−0.65)
COMP88	−12.418	−4.014	−4.610	29.663**	−0.820	−1.741	−75.005
	(−0.512)	(−0.29)	(−0.277)	(2.741)	(−0.344)	(−0.349)	(−0.677)
CMSA	−5054.369	−519.660	−3588.878	−149.027	11889**	−443.512	−229.346
	(−1.243)	(−0.497)	(−1.117)	(−0.369)	(21.003)	(−0.314)	(−0.164)
MSA	−1434.205	−156.640	−166.334	11.818	57.766	−1023.290	−568.628
	(−0.757)	(−0.165)	(−0.06)	(0.032)	(0.213)	(−0.721)	(−0.421)

F-VALUE	1.226	1.512	1.091	1.8	86.6	5.852	1.22
R-SQ	0.306	0.144	0.193	0.238	0.969	0.455	0.262
ADJ. R-SQ	0.056	0.049	0.016	0.106	0.958	0.378	0.047
N	35	91	51	62	35	73	41
CON. INDEX	21.228	19.335	23.378	19.6	23.107	20.332	25.089

NOTE 1: * = significance at the .10 level, ** = significance at the .05 level t statistics are in parentheses.

NOTE 2: a. metropolitan areas and non-metropolitan counties, b. municipalities.

Table 4.23 (continued)

DEP. VARIABLE: HIGH-TECH EMPLOYMENT GROWTH (TYG95)

VAR.	Communi-cations	Measurement and Testing Equipment		Software		Research, Develop-ment and Testing
		INTELLECTUAL-CAPITAL INDUSTRIES				
	b	a	b	a	b	b
INTERCEP	894.666	−453.659	2101.875	−2597.961	1225.943	−279.911
	(0.408)	(−0.125)	(0.956)	(−0.743)	(0.563)	(−0.386)
SP88	−162.916	29.733	−36.531	−94.532	87.174	−53.472*
	(−1.217)	(0.262)	(−0.576)	(−0.407)	(0.538)	(−1.791)
D88	−1358.796	−4642.496**	−2496.091*	−3161.974	−3792.022**	309.470
	(−1.241)	(−2.116)	(−1.912)	(−1.309)	(−3.461)	(0.854)
Y88	0.320*	−5.099**	1.011**	0.207**	−0.212**	2.075**
	(1.737)	(−6.217)	(2.18)	(2.082)	(−2.244)	(16.362)
MW88	−0.002	0.055	−0.019	0.165	0.100	−0.013
	(−0.031)	(0.499)	(−0.307)	(1.653)	(1.539)	(−0.805)
COLL	−35.078	256.162**	−2.667	203.248**	53.392*	2.869
	(−1.079)	(2.11)	(−0.084)	(2.554)	(1.93)	(0.288)
ACCESS	6.885	8.500*	4.834	2.206	−0.435	1.334
	(1.475)	(1.684)	(1.047)	(0.562)	(−0.108)	(1.107)
COMP88	−1.230	−145.704*	0.404	−0.445	−18.578	18.163
	(−0.119)	(−1.765)	(0.011)	(−0.061)	(−1.045)	(1.666)
CMSA	1470.237	57218**	1026.165	3233.224	−2038.963	647.148
	(1.209)	(9.695)	(0.871)	(0.931)	(−1.515)	(1.424)
MSA	−135.099	−1805.633	−376.872	−3131.973*	−1664.208	−175.098
	(−0.118)	(−1.214)	(−0.365)	(−1.937)	(−1.532)	(−0.467)
F-VALUE	1.354	18.649	1.866	9.85	2.144	34.878
R-SQ	0.147	0.808	0.117	0.816	0.211	0.865
ADJ. R-SQ	0.038	0.764	0.055	0.733	0.113	0.84
N	81	50	136	30	82	59
CON. INDEX	21.232	22.584	20.418	23.343	21.76	19.59

NOTE 1: * = significance at the .10 level, ** = significance at the .05 level t statistics are in parentheses.

NOTE 2: a. metropolitan areas and non-metropolitan counties, b. municipalities.

Table 4.23 (continued)

DEP. VARIABLE: HIGH-TECH EMPLOYMENT GROWTH (TYG95)

VAR.	VARIETY-BASED INDUSTRIES					
	Chemicals		Pharma-ceuticals and Biological Products	Material Handling Equipment	Other Manufacturing	
	a	b	a	b	a	b
INTERCEP	−1597.512	−778.609	502.572	−1.288	366.170	−422.332
	(−1.206)	(−1.076)	(1.139)	(−0.004)	(0.329)	(−0.573)
SP88	−681.718**	−94.143**	−6.603	−0.638	19.676	−117.514**
	(−4.999)	(−2.469)	(−0.848)	(−0.888)	(0.232)	(−2.047)
D88	1081.580	157.957	−208.961	−68.149	−2412.555**	840.516
	(1.108)	(0.339)	(−0.742)	(−0.365)	(−2.881)	(1.652)
Y88	4.239**	3.073**	−0.811**	−0.274	−0.031	2.619**
	(20.248)	(25.036)	(−4.053)	(−0.207)	(−0.323)	(33.121)
MW88	0.094**	0.046**	−0.008	0.012	0.043	0.010
	(3.329)	(2.845)	(−0.518)	(1.181)	(1.65)	(0.551)
COLL	−68.902	−7.631	−3.795	−2.185	95.595**	−5.188
	(−1.547)	(−0.575)	(−0.462)	(−0.325)	(2.785)	(−0.434)
ACCESS	−1.441	−0.212	−0.764	−0.463	0.290	−2.331*
	(−0.843)	(−0.165)	(−0.57)	(−0.656)	(0.2)	(−1.663)
COMP88	21.446	12.940	2.011	43.829**	−8.194	7.681
	(0.851)	(0.72)	(0.253)	(2.132)	(−0.217)	(0.917)
CMSA	13848.000**	239.219	168.788	−116.155	42663.000**	−569.526
	(7.912)	(0.67)	(0.485)	(−0.622)	(17.326)	(−1.597)
MSA	1506.571**	162.822	238.929	7.042	−503.394	97.877
	(2.477)	(0.527)	(0.869)	(0.041)	(−1.077)	(0.32)
F-VALUE	201.6	82.031	2.888	0.952	266.02	141.026
R-SQ	0.977	0.863	0.4	0.173	0.97	0.87
ADJ. R-SQ	0.973	0.853	0.261	−0.009	0.967	0.864
N	52	127	49	51	83	199
CON. INDEX	17.886	17.569	18.57	21.549	21.587	19.384

NOTE 1: * = significance at the .10 level, ** = significance at the .05 level t statistics are in parentheses.

NOTE 2: a. metropolitan areas and non-metropolitan counties, b. municipalities.

Table 4.24 The Determinants of the Degree of Specialization: Cross-sectional Analysis: by Product

DEP. VARIABLE: SPECIALIZATION (SP95)

VAR.	INNOVATION-BASED INDUSTRIES						
	Microelectronics		Computer and Peripherals	Medical Equipment	Aerospace/Aircraft		Military Equipment
	a	b	b	b	a	b	b
INTERCEP	4.902259**	7.49002**	4.458141*	19.66316**	6.487299**	8.615116**	22.77761**
	(2.47)	(5.317)	(1.832)	(3.481)	(3.164)	(4.671)	(5.591)
TYG95	0.158	0.139	0.063412	1.631	2.762**	0.029819	0.34
	(1.321)	(1.397)	(0.843)	(1.086)	(2.664)	(0.286)	(1.032)
Y88	0.054	0.017	−0.017	4.058	0.853**	0.013	0.136
	(0.647)	(0.207)	(−0.189)	(0.783)	(2.664)	(0.13)	(0.397)
MW88	−0.128	−0.189**	0.013738	0.035086	−0.155*	−0.08579	−0.48**
	(−1.693)	(−3.869)	(0.178)	(0.164)	(−1.731)	(−1.079)	(−2.713)
COLL	−0.00425	−0.02917	0.003114	−0.18715	0.056489	−0.08213*	−0.12752
	(−0.05)	(−1.024)	(0.098)	(−1.491)	(0.561)	(−1.779)	(−1.435)
ACCESS	0.002365	0.001098	−0.00788*	−0.01126	0.002042	−0.00319	−0.01424
	(0.61)	(0.344)	(−1.704)	(−0.879)	(0.342)	(−0.534)	(−1.431)
COMP88	0.016761	0.014249	−0.01671**	0.036964	0.007392	−0.00209	−0.38154*
	(1.156)	(1.131)	(−2.14)	(0.288)	(0.586)	(−0.504)	(−1.994)
CMSA	−1.20534	0.736106	−3.23097**	−11.326**	−35.7301**	−1.65899	−1.69461
	(−0.498)	(0.784)	(−2.107)	(−2.815)	(−2.837)	(−1.384)	(−0.64)
MSA	−1.44573	−0.82795	−1.70717	−10.4785**	−3.97616**	−2.98635**	−2.81347
	(−1.371)	(−0.964)	(−1.282)	(−2.803)	(−3.168)	(−2.545)	(−1.094)
F-VALUE	1.783	3.852	1.195	2.324	4.478	3.81	4.515
R-SQ	0.354	0.273	0.185	0.26	0.58	0.323	0.53
ADJ. R-SQ	0.156	0.202	0.03	0.148	0.45	0.238	0.413
N	35	91	51	62	35	73	41
CON. INDEX	14.967	16.037	20.655	15.746	23.9	16.623	18.665

NOTE 1: * = significance at the .10 level, ** = significance at the .05 level t statistics are in parentheses.

NOTE 2: a. metropolitan areas and non-metropolitan counties, b. municipalities.

NOTE 3: The coefficients for TYG95, Y88, and DMW were multiplied by 1,000 persons.

Table 4.24 (continued)

DEP. VARIABLE: SPECIALIZATION (SP95)

VAR.	Communications	Measurement and Testing Equipment		Software		Research, Development and Testing
	b	a	b	a	b	b
INTERCEP	5.188615**	4.079648**	6.410749**	3.499586	2.543605	3.936377
	(2.499)	(2.117)	(5.821)	(1.276)	(1.564)	(0.886)
TYG95	−0.00609	0.028889	0.039114	0.139	0.094658	0.415
	(−0.05)	(0.295)	(0.703)	(0.663)	(0.994)	(0.413)
Y88	0.238	0.002	−0.122	0.012262	−0.00304	−0.852
	(1.314)	(0.002)	(−0.426)	(0.114)	(−0.04)	(−0.393)
MW88	−0.02438	−0.08305	−0.05771	0.016056	−0.00204	0.171
	(−0.371)	(−1.222)	(−1.485)	(0.155)	(−0.036)	(1.426)
COLL	−0.03401	0.053126	−0.01776	−0.05603	0.058555**	−0.04462
	(−1.002)	(0.648)	(−0.897)	(−0.642)	(2.366)	(−0.66)
ACCESS	−0.0023	0.000932	−0.00288	0.000654	0.001054	−0.00648
	(−0.463)	(0.279)	(−0.985)	(0.167)	(0.306)	(−0.742)
COMP88	0.002842	−0.03537	−0.01437	−0.00032	−0.0024	−0.08408
	(0.264)	(−0.667)	(−0.62)	(−0.048)	(−0.161)	(−1.177)
CMSA	−1.98662	−2.56086	−2.50386**	−3.77632	−2.6221**	−5.26002
	(−1.558)	(−0.377)	(−3.363)	(−1.177)	(−2.367)	(−1.649)
MSA	−1.38773	−2.29724**	−2.69824**	−2.76469**	−3.21835**	−2.85449
	(−1.159)	(−2.495)	(−4.147)	(−2.348)	(−3.581)	(−1.057)
F-VALUE	1.009	2.246	5.675	1.567	2.525	0.818
R-SQ	0.101	0.305	0.263	0.374	0.217	0.116
ADJ. R-SQ	0.001	0.169	0.217	0.135	0.131	−0.026
N	81	50	136	30	82	59
CON. INDEX	18.436	19.309	16.053	19.226	18.699	16.158

NOTE 1: * = significance at the .10 level, ** = significance at the .05 level t statistics are in parentheses.

NOTE 2: a. metropolitan areas and non-metropolitan counties, b. municipalities.

NOTE 3: The coefficients for TYG95, Y88, and DMW were multiplied by 1,000 persons.

Table 4.24 (continued)

DEP. VARIABLE: SPECIALIZATION (SP95)

VAR.	Chemicals		Pharmaceuti- cals and Biological Products	Material Handling Equipment	Other Manufacturing	
	a	b	a	b	a	b
INTERCEP	1.864216	3.535175**	−7.21205	99.61608**	2.916759**	3.453551**
	(1.628)	(4.333)	(−0.429)	(4.26)	(4.431)	(7.922)
TYG95	0.181	0.221	−1.75	1.493	0.034627	0.101
	(1.208)	(1.614)	(−0.273)	(0.112)	(0.359)	(1.635)
Y88	−0.54	−0.658	−10.738	94.984	−0.04119	−0.294*
	(−0.926)	(−1.49)	(−1.238)	(0.91)	(−0.488)	(−1.732)
MW88	0.085747**	0.068474**	0.174	−2.124**	−0.05215**	−0.03879**
	(2.29)	(2.744)	(0.297)	(−2.475)	(−2.305)	(−2.482)
COLL	−0.1109*	−0.10666**	−0.07505	−1.1216**	0.014726	−0.02253**
	(−1.972)	(−5.609)	(−0.233)	(−2.112)	(0.469)	(−2.262)
ACCESS	0.001711	−0.00056	0.179022**	−0.01211	0.000317	−0.00018
	(0.799)	(−0.288)	(3.395)	(−0.198)	(0.251)	(−0.146)
COMP88	−0.0231	−0.0498*	−0.34198	−2.09965	−0.06211**	−0.00629
	(−0.773)	(−1.93)	(−1.522)	(−1.229)	(−2.221)	(−0.89)
CMSA	−3.89748	−1.06102**	9.535088	3.74906	−1.15403	−0.23512
	(−1.287)	(−2.016)	(0.688)	(0.252)	(−0.244)	(−0.771)
MSA	−0.83761	−0.79167*	−10.4441	−8.99869	−0.90883**	−0.22558
	(−1.284)	(−1.748)	(−0.951)	(−0.646)	(−2.636)	(−0.87)
F-VALUE	2.946	8.44	2.77	2.262	2.957	3.928
R-SQ	0.354	0.364	0.357	0.301	0.242	0.142
ADJ. R-SQ	0.234	0.321	0.228	0.168	0.16	0.106
N	52	127	49	51	83	199
CON. INDEX	14.637	13.164	15.989	16.475	17.988	13.338

NOTE 1: * = significance at the .10 level, ** = significance at the .05 level t statistics are in parentheses.

NOTE 2: a. metropolitan areas and non-metropolitan counties, b. municipalities.

NOTE 3: The coefficients for TYG95, Y88, and DMW were multiplied by 1,000 persons.

Table 4.25 The Determinants of the Degree of Diversity: Cross-sectional Analysis: by Product

DEP. VARIABLE: DIVERSITY (D95)

VAR.	INNOVATION-BASED INDUSTRIES						
	Microelectronics		Computer and Peripherals	Medical Equipment	Aerospace/Aircraft		Military Equipment
	a	b	b	b	a	b	b
INTERCEP	0.661908**	0.957682**	0.221141	0.861791**	0.749132**	0.882629**	1.064668**
	(3.195)	(6.389)	(0.855)	(5.625)	(3.68)	(6.023)	(5.203)
TYG95	−0.07377**	0.001129	0.006229	−0.03375	−0.06493	−0.00711	−0.03156
	(−2.313)	(0.099)	(0.563)	(−0.677)	(−0.503)	(−0.543)	(−1.488)
Y88	0.102*	−0.00699	−0.00291	−0.01889	−0.00125	−0.01618	−0.06574**
	(1.945)	(−0.583)	(−0.139)	(−0.128)	(−0.026)	(−1.347)	(−2.189)
MW88	0.002278	−0.00915*	0.016157*	0.005645	−0.01176	−0.00305	−0.00465
	(0.272)	(−1.763)	(1.968)	(0.974)	(−1.284)	(−0.483)	(−0.533)
TYGOM95	0.023031	0.00287	0.001945	0.002162	0.008862	0.001266	0.004346
	(1.385)	(0.971)	(0.591)	(0.789)	(0.304)	(0.241)	(1.189)
YOM88	−0.04761*	−0.01431**	−0.0146**	−0.01024**	−0.01434	−0.0029	0.000759
	(−2.005)	(−2.795)	(−2.146)	(−2.175)	(−0.37)	(−0.34)	(0.09)
COLL	−0.00988	−0.00498	−0.00211	−0.0079**	0.003255	−0.01264**	−0.00628
	(−1.126)	(−1.647)	(−0.621)	(−2.291)	(0.278)	(−3.168)	(−1.428)
ACCESS	0.000527	0.00013	−2.1E−05	−0.0005	0.001005*	0.000628	−4.6E−05
	(1.239)	(0.381)	(−0.042)	(−1.435)	(1.732)	(1.324)	(−0.094)
COMP88	0.000882	0.000921	0.00022	−0.00066	0.002733*	0.000915**	−0.04157**
	(0.584)	(0.689)	(0.265)	(−0.182)	(1.966)	(2.717)	(−3.381)
CMSA	−0.8368	−0.05703	−0.14065	−0.31495**	0.077017	−0.07362	−0.24241*
	(−0.414)	(−0.57)	(−0.844)	(−2.849)	(0.041)	(−0.777)	(−1.803)
MSA	−0.1873	−0.1167	−0.0846	−0.1927*	−0.4043**	−0.2823**	−0.3947**
	(−1.669)	(−1.282)	(−0.598)	(−1.895)	(−3.509)	(−3.03)	(−3.071)
F-VALUE	2.055	3.679	1.651	3.82	3.349	5.488	4.831
R-SQ	0.461	0.315	0.292	0.428	0.583	0.47	0.617
ADJ. R-SQ	0.237	0.229	0.115	0.316	0.409	0.384	0.489
N	35	91	51	62	35	73	41
CON. INDEX	40.8	16.587	21.698	16.407	73.447	17.358	20.05

NOTE 1: * = significance at the .10 level, ** = significance at the .05 level t statistics are in parentheses.

NOTE 2: a. metropolitan areas and non-metropolitan counties, b. municipalities.

NOTE 3: The coefficients for TYG95, Y88, DMW, TYGOM95, and YOM88 were multiplied by 1,000 persons.

Table 4.25 (continued)

DEP. VARIABLE: DIVERSITY (D95)

VAR.	Communi-cations	Measurement and Testing Equipment		Software		Research, Development and Testing
	b	a	b	a	b	b
INTERCEP	0.715452**	0.57267**	0.933123**	0.713019**	0.622048**	0.471915**
	(3.74)	(2.509)	(7.491)	(2.924)	(3.582)	(2.343)
TYG95	0.010575	−0.00282	0.001282	0.039016	−0.01535	0.031923
	(0.74)	(−0.08)	(0.121)	(0.493)	(−1.182)	(0.657)
Y88	−0.01723	−0.0576	−0.102	0.004286	−0.01622	−0.121
	(−0.988)	(−0.229)	(−0.968)	(0.031)	(−1.122)	(−1.228)
MW88	0.00329	0.010017	0.000229	0.004001	0.01412**	0.012877**
	(0.542)	(1.227)	(0.052)	(0.421)	(2.334)	(2.363)
TYGOM95	0.002644	−0.00188	0.009611	−0.01105	0.002823	0.00416
	(0.85)	(−0.099)	(1.235)	(−0.554)	(0.783)	(1.091)
YOM88	−0.01389**	−0.00289	−0.01371**	−0.01096	−0.01013**	−0.01213**
	(−2.732)	(−0.294)	(−2.265)	(−0.22)	(−2.212)	(−2.695)
COLL	−0.00435	−0.00528	−0.00781**	−0.00307	0.00204	−0.00538*
	(−1.386)	(−0.538)	(−3.455)	(−0.391)	(0.763)	(−1.74)
ACCESS	−0.00038	9.39E−05	−0.00028	8.28E−05	−0.00026	−7.2E−05
	(−0.826)	(0.214)	(−0.858)	(0.216)	(−0.693)	(−0.181)
COMP88	0.000986	−0.00694	−0.00197	0.001134	−0.00336**	−0.00071
	(0.996)	(−1.116)	(−0.754)	(1.658)	(−2.115)	(−0.208)
CMSA	−0.17895	0.363161	−0.12755	0.7306	−0.56238**	−0.21792
	(−1.525)	(0.358)	(−1.516)	(0.443)	(−4.713)	(−1.509)
MSA	−0.0794	−0.1761	−0.1406*	−0.3258**	−0.3799**	−0.0764
	(−0.72)	(−1.605)	(−1.913)	(−2.861)	(−3.942)	(−0.626)
F-VALUE	2.435	1.896	5.101	3.558	4.735	2.922
R-SQ	0.258	0.327	0.29	0.652	0.4	0.378
ADJ. R-SQ	0.152	0.155	0.233	0.469	0.316	0.249
N	81	50	136	30	82	59
CON. INDEX	19.226	37.082	16.631	108.304	19.337	17.163

NOTE 1: * = significance at the .10 level, ** = significance at the .05 level t statistics are in parentheses.

NOTE 2: a. metropolitan areas and non-metropolitan counties, b. municipalities.

NOTE 3: The coefficients for TYG95, Y88, DMW, TYGOM95, and YOM88 were multiplied by 1,000 persons.

Table 4.25 (continued)

DEP. VARIABLE: DIVERSITY (D95)

VAR.	VARIETY-BASED INDUSTRIES					
	Chemicals		Pharmaceutical	Material	Other Manufacturing	
	a	b	a	b	a	b
INTERCEP	0.55283**	0.929839**	0.25606	1.028006**	0.670471**	0.977621**
	(3.207)	(7.896)	(1.295)	(4.668)	(4.852)	(11.396)
TYG95	−0.00737	−0.01821	−0.08754	0.06988	−0.03179	0.007864
	(−0.268)	(−0.789)	(−0.678)	(0.468)	(−0.653)	(0.504)
Y88	−0.08057	0.015305	−0.08877	0.272	−0.00451	−0.0614
	(−0.824)	(0.232)	(−0.649)	(0.278)	(−0.049)	(−1.509)
MW88	0.00587	0.001334	0.008917	−0.00472	0.008528*	0.001033
	(1.028)	(0.368)	(1.29)	(−0.586)	(1.764)	(0.339)
TYGOM95	0.001389	0.006051	−1E−04	0.00312	0.00236	0.008071
	(0.102)	(1.273)	(−0.024)	(0.908)	(0.169)	(1.112)
YOM88	−0.01029	−0.01275**	−0.0055	−0.01256**	−0.00083	−0.01059**
	(−0.841)	(−2.722)	(−0.798)	(−2.355)	(−0.046)	(−1.654)
COLL	−0.00204	−0.01179**	−0.00595	−0.01156**	−0.00472	−0.00992**
	(−0.236)	(−4.162)	(−1.55)	(−2.251)	(−0.713)	(−5.006)
ACCESS	0.000571*	0.000175	0.001618**	6.19E−05	1.8E−06	−0.00017
	(1.773)	(0.618)	(2.608)	(0.11)	(0.006)	(−0.689)
COMP88	0.000728	−0.00336	−0.00051	−0.01332	0.004415	−0.00118
	(0.169)	(−0.874)	(−0.191)	(−0.779)	(0.751)	(−0.862)
CMSA	1.077911	−0.11885	0.044973	−0.08595	0.824704	−0.13334**
	(0.751)	(−1.563)	(0.276)	(−0.622)	(0.411)	(−2.243)
MSA	−0.3097**	−0.1324**	−0.1081	−0.1269	−0.2391**	−0.1041**
	(−3.27)	(−2.026)	(−0.836)	(−0.986)	(−3.22)	(−2.067)
F-VALUE	3.564	6.641	2.766	2.947	3.658	8.508
R-SQ	0.465	0.364	0.421	0.424	0.337	0.312
ADJ. R-SQ	0.335	0.309	0.269	0.28	0.245	0.275
N	52	127	49	51	83	199
CON. INDEX	31.551	13.702	16.689	18.321	53.758	14.163

NOTE 1: * = significance at the .10 level, ** = significance at the .05 level t statistics are in parentheses.

NOTE 2: a. metropolitan areas and non-metropolitan counties, b. municipalities.

NOTE 3: The coefficients for TYG95, Y88, DMW, TYGOM95, and YOM88 were multiplied by 1,000 persons.

2. Availability of skilled labor (COLL) was negatively and significantly associated with current specialization (SP95) of most variety-based industries, whereas this variable was positively and significantly associated with that of software. This reflects the relative importance of skilled labor in the software industry.

3. Among the other variables, availability of highly-skilled labor (COLL) was an important determinant of current diversity (D95) in most industries. Interestingly, in most industries, except for volatile industries such as aerospace/aircraft and military equipment, the initial level of manufacturing industries was one of the important sources of current diversity. This implies that high-tech development was not independent of the initial level of manufacturing industries.

4.4.3 The Effects of Dynamic Externalities on High-tech Employment Growth: Time-series Analysis

The empirical results for the simultaneous equation model for individual products now will be reported in this and the following section. This section deals with the results for equation (29), which states that current high-tech employment growth (YG95) is a function of yearly differences of specialization (DSP), yearly differences of diversity (DD), persistence (initial level) of high-tech employment (Y88), controlling for regional variables and dummies. The Hausman specification test reveals that in most products there is simultaneity between high-tech employment growth and dynamic externalities. However, in some products such as microelectronics, medical equipment, and research, development, and testing service, there is no such evidence. In these cases, OLS estimates are reported.

The regression results are laid out in Table 4.26 and the major findings are as follows:

1. Yearly differences in specialization (DSP) have positive effects on current high-tech growth. This evidence supports the MAR view. The positive effects were significant for microelectronics and medical equipment. But, negative effects also were observed as in the cross-sectional analysis. The negative effects were significant for chemicals. Figure 4.6 clearly shows the lagged effects of specialization on current high-tech employment in these industries. The lag pattern seems to be different, depending on each individual product. It seems that chemicals have a lag structure longer than 6 years, whereas other industries have a relatively short time-lag. In addition, somewhat surprisingly, in some products such as aerospace/aircraft and software, there were negative effects which became prominent, reflecting decreasing economies of scale in these industries.

Table 4.26 The Effects of Historical Conditions on High-tech Employment Growth: Time-series Analysis: by Product

DEP. VARIABLE: HIGH-TECH EMPLOYMENT GROWTH (YG95)

VAR.	INNOVATION-BASED INDUSTRIES					
	Microelectronics		Medical Equipment		Aerospace/Aircraft	
	COEFF.	T STAT.	COEFF.	T STAT.	COEFF.	T STAT.
INTERCEP	−356.005	−0.613	−451.824**	−2.695	244.5113	0.15
DSP95	117.2347*	1.775	76.84072**	5.624	58.04488	0.404
DSP94	−102.601	−1.41	4.49762	0.731	−179.615	−1.296
DSP93	21.01876	0.289	2.770128	0.437	−15.5777	−0.129
DSP92	34.55309	0.644	0.928591	0.165	−44.7004	−0.24
DSP91	27.24071	0.296	39.39487**	2.838	−237.882	−0.799
DSP90	21.16382	0.257	8.76097**	2.039	−37.3161	−0.272
DSP89	160.7694	0.974	8.194012**	2.468	5.949533	0.032
DD95	58.0444	0.059	−80.4952	−0.305	1589.013	0.719
DD94	1114.546	1.224	15.75287	0.06	167.8341	0.075
DD93	696.9524	0.883	−375.229*	−1.8	2126.337	0.884
DD92	716.8079	0.932	589.7755**	2.11	1473.18	0.787
DD91	33.82026	0.047	−95.0151	−0.323	2267.37	1.071
DD90	903.0488	1.531	−98.7752	−0.615	2016.83	1.594
DD89	399.8685	0.805	111.2921	0.754	1245.622	1.034
Y88	0.143033**	4.637	0.142548	0.885	0.031306	0.632
MW95	0.023288	1.456	0.005626	1.145	0.027229	0.625
COLL	2.535124	0.237	9.145918**	2.72	3.304738	0.103
ACCESS	−0.10732	−0.092	0.435669	1.296	−0.04319	−0.009
COMP88	−0.70262	−0.139	15.8715**	4.378	−2.11411	−0.718
CMSA	−468.222	−1.198	−77.1161	−0.705	−70.2135	−0.09
MSA	−3.48121	−0.01	−129.007	−1.196	−596.236	−0.699
F-VALUE	2.082		8.245		0.461	
R-SQ	0.391		0.812		0.165	
ADJ. R-SQ	0.203		0.714		−0.193	
N	90		62		71	
CON. INDEX	20.36		22.348		25.628	

NOTE: * = significance at the .10 level, ** = significance at the .05 level.

Table 4.26 (continued)

DEP. VARIABLE: HIGH-TECH EMPLOYMENT GROWTH (YG95)

VAR.	INTELLECTUAL-CAPITAL INDUSTRIES							
	Communications		Measurement and Testing Equipment		Software		Research, Development and Testing	
	COEFF.	T STAT.	COEFF.	T STAT.	COEFF.	T STAT.	COEFF.	T STAT.
INTERCEP	−1047.55	−0.824	65.99537	0.629	719.3833	0.799	−62.2581	−0.122
DSP95	153.1005	0.208	9.506012	0.363	−186.895	−0.905	17.87457	1.04
DSP94	31.16831	0.195	8.933979	0.875	−118.167	−0.483	−62.6879	−0.523
DSP93	−42.2407	−0.507	−1.03575	−0.123	−83.7202	−1.021	15.73678	0.413
DSP92	−56.7819	−0.53	−1.63305	−0.314	−128.622	−0.748	6.190267	0.401
DSP91	−14.7027	−0.096	−4.75296	−0.809	−95.4747	−0.799	45.63845	1.084
DSP90	−1.10046	−0.015	−2.04266	−0.573	20.93437	0.217	13.72559	0.838
DSP89	66.53979	0.444	−0.08394	−0.017	−64.6662	−0.653	17.33117	1.007
DD95	2492.049	0.662	−133.252	−0.297	6027.329	1.393	−74.2439	−0.09
DD94	714.0535	0.372	−102.174	−0.351	1167.188	1.012	−562.523	−0.638
DD93	−326.29	−0.301	54.60626	0.39	3034.128	1.345	−408.977	−0.958
DD92	554.6926	0.341	−25.4243	−0.218	2280.764	1.351	−665.617	−1.138
DD91	1231.482	1.111	55.44861	0.371	1200.818	1.007	129.7684	0.334
DD90	457.434	0.589	−22.0323	−0.204	1460.879	1.389	−47.8258	−0.144
DD89	−397.392	−0.619	−20.9591	−0.276	205.1012	0.328	17.526	−0.061
Y88	0.18798**	3.076	0.017087	0.87	0.29173**	6.761	0.1484**	2.595
MW95	0.002084	0.113	−0.00165	−0.69	−0.00781	−0.372	−0.00641	−0.814
COLL	−5.42474	−0.304	0.463386	0.321	−10.3013	−0.711	5.269253	1.111
ACCESS	8.38965**	2.063	−0.08199	−0.376	1.163237	0.621	0.322197	0.478
COMP88	−1.68769	−0.43	−1.39199	−0.742	3.521078	0.449	7.529486	1.494
CMSA	1250.108*	1.751	3.64176	0.067	543.6824	0.797	223.0086	0.67
MSA	10.13806	0.021	27.54422	0.567	−36.7356	−0.068	24.33791	0.086
F-VALUE	3.003		0.311		4.704		0.863	
R-SQ	0.517		0.055		0.63		0.335	
ADJ. R-SQ	0.345		−0.122		0.496		−0.053	
N	81		134		80		58	
CON. INDEX								

NOTE: * = significance at the .10 level, ** = significance at the .05 level.

Table 4.26 (continued)

DEP. VARIABLE: HIGH-TECH EMPLOYMENT GROWTH(YG95)

	VARIETY-BASED INDUSTRIES							
VAR.	Chemicals a		Chemicals b		Other Manufacturing a		Other Manufacturing b	
	COEFF.	T STAT.	COEFF.	T STAT.	COEFF.	T STAT.	COEFF.	T STAT.
INTERCEP	286.3816	0.521	496.9744	0.905	−1336.71*	−1.88	−262.275	−0.664
DSP95	−99.3342	−0.245	−93.6386	−0.218	44.62636	0.069	16.21747	0.048
DSP94	−1796.2**	−4.534	−89.4849	−0.986	101.72	0.655	−10.7829	−0.139
DSP93	−378.22**	−3.141	−91.7433	−1.308	10.09116	0.097	−48.4306	−0.7
DSP92	−779.36**	−2.78	−43.4282	−0.356	24.64464	0.249	3.414582	0.045
DSP91	−252.83**	−3.328	−84.0085	−1.304	−59.3371	−0.381	55.32373	0.827
DSP90	−284.47**	−3	−213.891	−1.316	−54.3112	−0.388	26.37955	0.548
DSP89	−289.84**	−5.085	−50.1504	−1.142	−34.1973	−0.386	−24.3558	−0.444
DD95	1015.525	0.902	2513.998	0.78	1424.865	0.354	4275.69*	1.838
DD94	924.2374	0.632	418.8069	0.349	821.8096	0.608	3351.96*	1.688
DD93	1680.7*	1.915	904.109	0.74	252.988	0.271	904.8752	1.404
DD92	214.6077	0.245	294.1532	0.247	495.083	0.42	483.5936	0.79
DD91	9.476947	0.014	486.3976	0.682	757.1447	0.647	1052.353	1.615
DD90	2.714383	0.006	1042.666	1.571	889.6202	0.931	828.2121	1.399
DD89	−128.833	−0.264	559.9275	1.385	460.6052	0.611	8.758657	0.022
Y88	−0.5767**	−8.442	0.041584	0.493	0.3566**	4.945	0.1131**	2.175
MW95	0.01578	1.475	0.0037	0.384	0.02641	1.641	0.011051	1.196
COLL	−27.3893	−1.338	−2.9575	−0.262	78.032**	3.014	6.447684	0.773
ACCESS	−1.19822	−1.385	−0.93208	−0.906	−0.58302	−0.518	0.279612	0.29
COMP88	−29.58**	−2.919	4.56464	0.329	−8.1155	−0.214	2.283788	0.401
CMSA	7319.5**	14.495	−113.284	−0.376	−8651.1**	−4.733	−157.518	−0.639
MSA	543.79**	2.483	18.31773	0.084	−155.964	−0.444	128.2765	0.634
F-VALUE	14.999		0.615		2.641		0.831	
R-SQ	0.924		0.115		0.493		0.091	
ADJ. R-SQ	0.862		−0.072		0.306		−0.019	
N	48		121		79		196	
CON. INDEX								

NOTE 1: * = significance at the .10 level, ** = significance at the .05 level.
NOTE 2: a. metropolitan areas and non-metropolitan counties, b. municipalities.

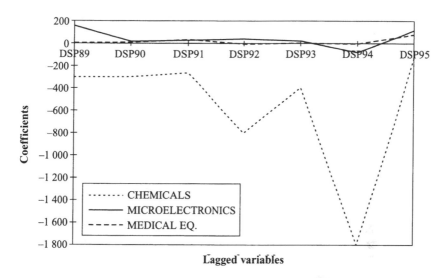

Figure 4.6 The Lagged Effects of Specialization on High-tech Employment Growth: by Product

NOTE: Sample population is the MAs and Counties for chemicals and the municipalities for microelectronics and medical equipment.

2. As in previous tests, the lagged diversity variables have positive signs: the lack of diversity is positively associated with high-tech employment growth. However, in some industries such as medical equipment, measurement and controlling equipment, and research, development and testing, Jacobs effects are observed. The evidence shows that current high-tech employment growth also depends on past diversity, although the negative effects of yearly differences of the diversity variable (DD) were not strong, as shown in Figure 4.7.

3. A very high degree of persistence was still observed in many industries, despite strong evidence for relocation. The exception was the chemical industry, where the sign for this variable was significantly negative. As shown in other tests, chemicals evidence decreasing returns to scale.

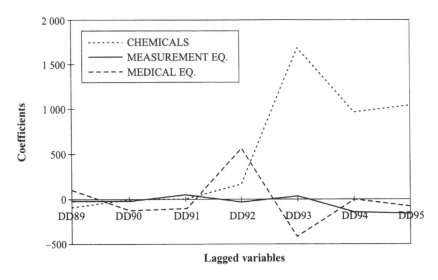

Figure 4.7 The Lagged Effects of Diversity on High-tech Employment Growth: by Product

NOTE: Sample population is the MAs and Counties for chemicals and the municipalities for measurement equipment and medical equipment.

4.4.4 The Determinants of the Degree of Dynamic Externalities: Time-series Analysis

I now proceed to equations (30) and (31). Current specialization (SP95) and current diversity (D95) are assumed to depend on total high-tech employment (TYG95) and persistence (initial level) of high-tech employment (Y88), controlling for regional variables and dummies. Unlike cross-sectional analysis, yearly differences of manufacturing wages (DMW) are included as control variables.

Tables 4.27 and 4.28 show the regression results. The major findings are as follows:

1. Current specialization (SP95) was not dependent upon the initial employment level (Y88) and growth (TYG95) of individual products. This is also true for diversity. This finding is consistent with the cross-sectional analysis.

Table 4.27 The Determinants of the Degree of Specialization: Time-series Analysis: by Product

DEP. VARIABLE: SPECIALIZATION (SP95)

VAR.	INNOVATION-BASED INDUSTRIES					
	Microelectronics		Medical Equipment		Aerospace/Aircraft	
	COEFF.	T STAT.	COEFF.	T STAT.	COEFF.	T STAT.
INTERCEP	4.2951**	2.951	12.1136*	1.977	7.2829**	3.517
TYG95	0.0727	0.692	1.4000	0.944	0.0307	0.293
Y88	0.0120	0.141	1.3910	0.269	0.0167	0.163
DMW95	−0.2660*	−1.888	0.9630*	1.753	−0.1310	−0.731
DMW94	−0.4610	−1.605	0.5950	0.554	0.3240	0.887
DMW93	0.1620	0.454	2.5010	1.537	−0.0976	−0.269
DMW92	−0.7430**	−2.632	1.4440	1.378	−0.3650	−1.066
DMW91	−0.2990	−1.059	1.6490*	1.718	−0.1730	−0.447
DMW90	0.4450*	1.677	1.4770	1.443	0.1230	0.336
DMW89	0.2310	1.015	−0.5890	−0.554	0.3730	0.997
COLL	−0.0127	−0.411	−0.2738**	−2.088	−0.0942*	−1.874
ACCESS	0.0043	1.225	0.0026	0.191	−0.0020	−0.268
COMP88	0.0066	0.481	0.0723	0.553	−0.0012	−0.293
CMSA	−0.6629	−0.624	−13.5262**	−3.254	−1.7088	−1.375
MSA	−1.3692	−1.457	−11.1781**	−2.822	−3.5502**	−2.948
F-VALUE	1.987		1.895		2.183	
R-SQ	0.271		0.361		0.353	
ADJ. R-SQ	0.134		0.17		0.191	
N	90		62		71	
CON. INDEX	16.413		16.744		16.817	

NOTE 1: * = significance at the .10 level, ** = significance at the .05 level.
NOTE 2: The coefficients for TYG95, Y88, and DMW were multiplied by 1,000 persons.

Table 4.27 (continued)

DEP. VARIABLE: SPECIALIZATION (SP95)

VAR.	INTELLECTUAL-CAPITAL INDUSTRIES							
	Communications		Measurement and Testing Equipment		Software		Research, Development and Testing	
	COEFF.	T STAT.	COEFF.	T STAT.	COEFF.	T STAT.	COEFF.	T STAT.
INTERCEP	4.8509**	3.225	5.0043**	5.078	2.4147	1.66	2.1889	0.479
TYG95	−0.0652	−0.395	−0.0988	−1.417	0.0827	0.616	0.6050	0.685
Y88	0.2990*	1.77	−0.0128	−0.043	−0.0095	−0.128	−1.1690	−0.609
DMW95	−0.2050	−1.482	0.0098	0.148	0.0396	0.318	0.1160	0.321

DMW94	-0.9200**	-2.342	0.0092	0.049	0.4630*	1.836	0.6820	1.116
DMW93	0.2980	0.797	0.2130	1.093	0.2740	0.902	2.4020**	3.173
DMW92	-0.7050**	-2.424	-0.2840	-1.58	0.1760	0.779	-0.2350	-0.348
DMW91	-0.0662	-0.255	-0.0921	-0.579	-0.0023	-0.015	-0.0132	-0.022
DMW90	0.4530	1.404	0.0674	0.389	-0.3910	-1.363	1.5290*	1.928
DMW89	-0.3900	-1.41	0.3020*	1.754	-0.1690	-0.672	-1.4330**	-2.684
COLL	-0.0084	-0.268	-0.0114	-0.561	0.0666**	2.669	0.0010	0.018
ACCESS	-0.0015	-0.358	-0.0001	-0.041	0.0007	0.197	-0.0096	-1.176
COMP88	0.0060	0.607	-0.0195	-0.787	-0.0059	-0.388	-0.1265*	-1.936
CMSA	-1.8798	-1.468	-2.8195**	-3.951	-3.1242**	-2.496	-5.2588	-1.515
MSA	-0.4445	-0.361	-3.3726**	-5.107	-3.5174**	-3.245	0.2682	0.089
F-VALUE	2.383		3.272		1.403		1.63	
R-SQ	0.336		0.278		0.232		0.347	
ADJ. R-SQ	0.195		0.193		0.067		0.134	
N	81		134		80		58	
CON. INDEX			21.265					

NOTE 1: * = significance at the .10 level, ** = significance at the .05 level.
NOTE 2: The coefficients for TYG95, Y88, and DMW were multiplied by 1,000 persons.

Table 4.27 (continued)

DEP. VARIABLE: SPECIALIZATION (SP95)

	VARIETY-BASED INDUSTRIES							
VAR.	Chemicals a		Chemicals b		Other Manufacturing a		Other Manufacturing b	
	COEFF.	T STAT.	COEFF.	T STAT.	COEFF.	T STAT.	COEFF.	T STAT.
INTERCEP	2.3459**	2.176	3.6939**	3.973	2.1258**	3.346	2.9327**	7.39
TYG95	0.3170**	2.06	0.2170	1.042	-0.0897	-0.862	-0.0645	-0.684
Y88	-0.9330	-1.561	-0.6350	-0.987	-0.0370	-0.445	0.1240	0.497
DMW95	-0.0399	-0.522	0.0744	1.368	-0.0403	-0.936	-0.0130	-0.402
DMW94	0.2320	1.5	0.1680	1.232	-0.0420	-0.527	0.0360	0.489
DMW93	0.2800*	1.751	0.1550	1.094	0.0043	0.047	-0.0185	-0.212
DMW92	0.2140	1.516	0.1020	0.878	-0.1550*	-1.709	-0.0731	-0.975
DMW91	-0.0745	-0.452	0.1090	0.802	-0.1610**	-1.916	-0.0833	-1.266
DMW90	0.1890	1.021	0.2180	1.326	-0.2450**	-2.678	-0.0855	-1.079
DMW89	0.2640	1.32	0.1190	0.826	-0.2700**	-2.683	-0.1330*	-1.84
COLL	-0.0841	-1.505	-0.1087**	-5.198	0.0471	1.439	-0.0197*	-1.884
ACCESS	0.0010	0.442	-0.0009	-0.361	0.0001	0.075	-0.0006	-0.408
COMP88	0.0243	0.841	-0.0259	-0.902	-0.0543*	-1.802	-0.0085	-1.154
CMSA	-6.3543*	-2.023	-0.3287	-0.581	3.7986	0.748	-0.5022*	-1.659
MSA	-0.7530	-1.123	-0.2848	-0.567	-0.8956**	-2.675	-0.3297	-1.222

F-VALUE	2.296	3.853	2.206	1.939
R-SQ	0.493	0.337	0.326	0.13
ADJ. R-SQ	0.279	0.25	0.178	0.063
N	48	121	79	196
CON. INDEX				

NOTE 1: * = significance at the .10 level, ** = significance at the .05 level.
NOTE 2: a. metropolitan areas and non-metropolitan counties, b. municipalities.
NOTE 3: The coefficients for TYG95, Y88, and DMW were multiplied by 1,000 persons.

Table 4.28 The Determinants of the Degree of Diversity: Time-series Analysis: by Product

DEP. VARIABLE: DIVERSITY (D95)

VAR.	INNOVATION-BASED INDUSTRIES					
	Microelectronics		Medical Equipment		Aerospace/Aircraft	
	COEFF.	T STAT.	COEFF.	T STAT.	COEFF.	T STAT.
INTERCEP	0.8994**	6.207	0.9551**	5.525	0.8330**	5.037
TYG95	−0.0019	−0.169	−0.0381	−0.736	−0.0045	−0.339
Y88	−0.0052	−0.444	−0.0017	−0.011	−0.0137	−1.122
DMW95	−0.0271*	−1.938	0.0065	0.421	0.0111	0.778
DMW94	−0.0335	−1.176	0.0016	0.054	0.0055	0.19
DMW93	0.0098	0.277	0.0123	0.268	0.0267	0.939
DMW92	−0.0763**	−2.73	−0.0301	−1.011	−0.0220	−0.823
DMW91	−0.0435	−1.551	0.0033	0.121	−0.0299	−0.972
DMW90	0.0394	1.484	0.0228	0.783	0.0177	0.615
DMW89	−0.0158	−0.702	0.0168	0.551	0.0069	0.236
TYGOM95	0.0020	0.685	0.0018	0.62	0.0013	0.237
YOM88	−0.0126**	−2.516	−0.0105**	−2.075	−0.0045	−0.53
COLL	−0.0035	−1.142	−0.0071*	−1.911	−0.0135**	−3.222
ACCESS	0.0001	0.37	−0.0005	−1.32	0.0008	1.385
COMP88	0.0002	0.153	−0.0015	−0.395	0.0009**	2.719
CMSA	−0.1180	−1.117	−0.2734**	−2.305	−0.1064	−1.094
MSA	−0.1072	−1.152	−0.1909*	−1.723	−0.3308**	−3.499
F-VALUE	3.102		2.585		3.453	
R-SQ	0.405		0.479		0.506	
ADJ. R-SQ	0.274		0.294		0.359	
N	90		62		71	
CON. INDEX	16.848		17.472		17.68	

NOTE 1: * = significance at the .10 level, ** = significance at the .05 level.
NOTE 2: The coefficients for TYG95, Y88, DMW, TYGOM95, and YOM88 were multiplied by 1,000 persons.

Table 4.28 (continued)

DEP. VARIABLE: DIVERSITY (D95)

	INTELLECTUAL-CAPITAL INDUSTRIES							
VAR.	Communications		Measurement and Testing Equipment		Software		Research Development and Testing	
	COEFF.	T STAT.	COEFF.	T STAT.	COEFF.	T STAT.	COEFF.	T STAT
INTERCEP	0.6932**	3.908	0.8989**	6.611	0.9675**	5.178	0.4485*	1.932
TYG95	0.0868	1.629	−0.0835	−1.186	0.0410	1.049	0.0365	0.758
Y88	−0.0088	−0.438	−0.6460	−1.378	0.0168	0.629	−0.1450	−1.491
DMW95	−0.0037	−0.225	−0.0059	−0.647	0.0095	0.593	0.0015	0.079
DMW94	−0.0289	−0.61	0.0032	0.116	0.0375	1.193	0.0729**	2.347
DMW93	0.0499	1.103	0.0266	0.943	0.0055	0.136	0.0915**	2.373
DMW92	−0.0518	−1.526	−0.0234	−0.911	0.0029	0.102	0.0000	−0.001
DMW91	0.0156	0.499	−0.0100	−0.434	0.0266	1.334	0.0311	1.036
DMW90	0.0873**	2.258	0.0252	1.058	−0.0552	−1.372	0.0543	1.284
DMW89	−0.0408	−1.25	−0.0129	−0.547	0.0122	0.386	−0.0743**	−2.731
TYGOM95	−0.0016	−0.385	0.0472	1.432	−0.0029	−0.505	0.0034	0.885
YOM88	−0.0236**	−2.39	0.0231	0.759	−0.0201**	−2.664	−0.0102**	−2.202
COLL	−0.0016	−0.407	−0.0088**	−2.776	−0.0012	−0.318	−0.0037	−1.212
ACCESS	−0.0012*	−1.84	0.0000	−0.094	−0.0003	−0.702	−0.0003	−0.617
COMP88	0.0012	1.055	−0.0020	−0.589	−0.0033*	−1.733	−0.0012	−0.332
CMSA	−0.2212	−1.402	−0.0807	−0.745	−0.4184**	−2.523	−0.1133	−0.644
MSA	0.1255	0.843	−0.1381	−1.529	−0.3471**	−2.527	0.1097	0.721
F-VALUE	1.922		2.572		2.218		2.418	
R-SQ	0.325		0.26		0.36		0.486	
ADJ. R-SQ	0.156		0.159		0.198		0.285	
N	81		134		80		58	
CON. INDEX							22.111	

NOTE 1: * = significance at the .10 level, ** = significance at the .05 level.
NOTE 2: The coefficients for TYG95, Y88, DMW, TYGOM95, and YOM88 were multiplied by 1,000 persons.

Table 4.28 (continued)

DEP. VARIABLE: DIVERSITY (D95)

VAR.	VARIETY-BASED INDUSTRIES							
	Chemicals a		Chemicals b		Other Manufacturing a		Other Manufacturing b	
	COEFF.	T STAT.	COEFF.	T STAT.	COEFF.	T STAT.	COEFF.	T STAT.
INTERCEP	0.6692**	4.127	0.8547**	6.353	0.6364**	3.925	1.0417**	9.229
TYG95	0.0207	0.782	0.0010	0.024	−0.3230**	−2.279	−0.1930**	−2.726
Y88	−0.1630*	−1.747	−0.0347	−0.3	−0.3640*	−1.929	0.2790**	2.198
DMW95	−0.0226**	−2.122	−0.0009	−0.121	−0.0101	−0.924	−0.0100	−1.181
DMW94	−0.0193	−0.912	−0.0015	−0.076	0.0302	1.493	0.0219	1.162
DMW93	0.0684**	3.147	0.0388*	1.911	0.0714**	3.06	0.0511**	2.276
DMW92	0.0171	0.889	0.0124	0.735	−0.0090	−0.345	0.0094	0.487
DMW91	−0.0420*	−1.873	0.0041	0.207	−0.0223	−0.968	0.0051	0.306
DMW90	−0.0151	−0.601	0.0090	0.375	−0.0183	−0.759	0.0312	1.516
DMW89	0.0161	0.593	0.0183	0.876	−0.0247	−0.953	−0.0204	−1.097
TYGOM95	0.0044	0.352	0.0057	1	0.0747**	2.123	0.0662**	3.018
YOM88	−0.0096	−0.884	−0.0145**	−2.899	0.0784*	1.952	−0.0102	−1.176
COLL	0.0044	0.528	−0.0113**	−3.545	0.0059	0.681	−0.0128**	−4.254
ACCESS	0.0005	1.334	0.0003	0.8	0.0005	1.096	−0.0008*	−1.936
COMP88	0.0046	1.136	−0.0015	−0.357	0.0040	0.504	−0.0011	−0.598
CMSA	0.1836	0.141	−0.1097	−1.339	11.3431**	2.151	−0.2530**	−2.997
MSA	−0.3667**	−3.745	−0.1264*	−1.743	−0.2363**	−2.683	−0.0774	−1.12
F-VALUE	3.294		3.706		2.443		3.785	
R-SQ	0.63		0.363		0.387		0.253	
ADJ. R-SQ	0.439		0.265		0.228		0.186	
N	48		121		79		196	
CON. INDEX								

NOTE 1: * = significance at the .10 level, ** = significance at the .05 level.
NOTE 2: a. metropolitan areas and non-metropolitan counties, b. municipalities.
NOTE 3: The coefficients for TYG95, Y88, DMW, TYGOM95, and YOM88 were multiplied by 1,000 persons.

2. Wage effects (DMW) on current specialization (SP95) seem to differ by product. For some products, such as microelectronics, communications and other high-tech manufacturing, wage differentials had negative signs, and persisted for a few years or longer: wage increases reduced the concentrations of high-tech industries. On the other hand, for other industries, such as medical equipment, software, and research, development, and testing service, the opposite is true: wage increases raised the concentrations of these industries. This result seems to be related to the importance of human capital in each industry. Availability of

highly-skilled labor (COLL) was positively significant only for the concentrations of software industry.

3. The important sources of current diversity (D95) for individual products were availability of highly-skilled labor (COLL) and the initial employment level of other manufacturing industries (YOM88). The effects of the wage variable on current diversity differ by product. For some industries, such as chemicals and microelectronics, the sign of this variable was significantly negative, whereas for other industries such as research, development, and testing service, that was significantly positive.

In summary, the test results for individual products support hypothesis 1. As expected, however, the effects of dynamic externalities on high-tech employment growth differ by product. Overall, specialization effects are not readily observable, unlike the previous tests. This seems to be related to firm size, as noted. On the other hand, Jacobs effects was observed for many industries, although the effects of yearly differences of diversity (DD) were not strong.

4.5 The Effects of Local Competition on Regional High-tech Employment Growth

4.5.1 Direct Effects of Local Competition on Regional High-tech Employment Growth

Previous test results show that there is no evidence that local competition is directly related to high-tech employment growth. Controlling for firm size and organizational type does not affect this relationship. However, for some products, the local competition variable (COMP88) was significant. Table 4.29 summarizes the test results. The third column shows the coefficients of the local competition variable (COMP88) for the cross-sectional analysis, whereas the fourth column shows the coefficients of this variable for the time-series analysis.

For some products, such as measurement and testing equipment and chemicals, this variable had negative effects on employment growth. But for medical equipment/devices and material handling equipment, the relationship was positive. The pattern is not closely related to broad classification by non-production occupation, the portions of highly-skilled labor, and the portions of diversification. This pattern is unlikely to be related to firm size. Another possible interpretation is that local competition might affect employment growth differentially, depending on the cooperational structure within industries.

Table 4.29 Direct Effects of Local Competition by Product

INDUSTRIES	SAMPLE POP.	TYG95	Y95
INNOVATION-BASED INDUSTRIES			
Microelectronics	a	−12.418	.
		(−0.512)	
	b	−4.014	−0.703
		(−0.29)	(−0.139)
Computer Systems and Peripherals/Accessories	b	−4.610	.
		(−0.277)	
Medical Equipment/Devices	b	29.663**	15.872**
		(2.741)	(4.378)
Aerospace/Aircraft and Equipment	a	−0.820	.
		(−0.344)	
	b	−1.741	−2.114107
		(−0.349)	(−0.718)
Military Equipment/Services	b	−75.005	.
		(−0.677)	
INTELLECTUAL-CAPITAL INDUSTRIES			
Communications Equipment/Services	b	−1.230	−1.688
		(−0.119)	(−0.43)
Measurement and Testing Equipment	a	−145.704*	.
		(−1.765)	
	b	0.404	−1.392
		(0.011)	(−0.742)
Software Development/Services	a	−0.445	.
		(−0.061)	
	b	−18.578	3.521078
		(−1.045)	(0.449)
Research, Development and Testing Services	b	18.163	7.529486
		(1.666)	(1.494)
VARIETY-BASED INDUSTRIES			
Chemicals	a	21.446	−29.576**
		(0.851)	(−2.919)
	b	12.940	4.565
		(0.72)	(0.329)
Pharmaceuticals and Biological Products	b	2.011	.
		(0.253)	
Material Handling Equipment	b	43.829**	.
		(2.132)	
Other High-tech Manufacturing Industries	a	−8.194	−8.115
		(−0.217)	(−0.214)
	b	7.681	2.284
		(0.917)	(0.401)

NOTE 1: * = significance at the .10 level, ** = significance at the .05 level (t statistics are in parentheses).

NOTE 2: a. metropolitan areas & non-metropolitan counties, b. municipalities.

4.5.2 Indirect Effects of Local Competition on Regional High-tech Employment Growth via Dynamic Externalities

Local competition is more likely to affect the level of specialization and diversity. In turn, these dynamic externalities seem to affect high-tech employment growth. By product, there is a great deal of variation. Table 4.30 summarizes the regression results. The third and the fourth columns show the coefficients of the local competition variable (COMP88) on specialization, whereas the fifth and the sixth columns show the coefficients of this variable on diversity. There are some important findings:

1. Local competition has negative impacts on own current specialization. For such products as computer and peripherals, chemicals, military equipment, and other manufacturing industries, the coefficients of this variable were negatively significant. This implies that when there is local competition, some high-tech industries tend to diversify their products.
2. Local competition affected current diversity either negatively or positively. In particular, whereas for aircraft/aerospace industry, there are positive effects, the negative effects are significant for software and military industries. One possible interpretation is that local competition facilitates the diversity of these industries. An alternative interpretation of this finding is that smaller firms encourage diversity. Although smaller firms grew faster for the period of 1988 to 1995, as shown later, military industry did not grew as fast during that period.

These test results for individual products support hypothesis 3. Although the effects of local competition were not strong, for some products, there is evidence that local competition encouraged their own employment growth. This is in favor of the Porter's and Jacobs' views, and also is consistent with the findings of Glaeser et al. (1991). Another important point is that local competition has indirect effects on employment growth by affecting dynamic externalities, especially diversity.

4.6 The Effects of High-tech Employment on Regional Economic Growth

Finally, I report on the correlation and multiple regression analyses of the relationship between initial high-tech employment and regional economic growth. First, I briefly describe trends for high-tech employment in the state of Texas between 1988 and 1995, with reference to employed civilian labor force. Then, I discuss the solutions to equations (32) and (33) in section 4.5.2. Lastly, I discuss the solutions to equations (34), (35) and (36) in section 4.5.3.

Table 4.30 Indirect Effects of Local Competition by Product

INDUSTRIES	SAMPLE POP.	SP95		D95	
		CROSS-SECTIONAL	TIME-SERIES	CROSS-SECTIONAL	TIME-SERIES
INNOVATION-BASED INDUSTRIES					
Microelectronics	a	0.017	.	0.0009	.
		(1.156)		(0.584)	
	b	0.014	0.0066	0.0009	0.0002
		(1.131)	(0.481)	(0.689)	(0.153)
Computer Systems and Peripherals/Accessories	b	−0.0167**	.	0.0002	.
		(−2.14)		(0.265)	
Medical Equipment/Devices	b	0.037	0.0723	−0.0007	−0.0015
		(0.288)	(0.553)	(−0.182)	(−0.395)
Aerospace/Aircraft and Equipment	a	0.007	.	0.0027*	.
		(0.586)		(1.966)	
	b	−0.002	−0.0012	0.0009**	0.0009**
		(−0.504)	(−0.293)	(2.717)	(2.719)
Military Equipment/Services	b	−0.382*	.	−0.042**	.
		(−1.994)		(−3.381)	
INTELLECTUAL-CAPITAL INDUSTRIES					
Communications Equipment/Services	b	0.003	0.006	0.0010	0.0012
		(0.264)	(0.607)	(0.996)	(1.055)
Measurement and Testing Equipment	a	−0.035	.	−0.0069	.
		(−0.667)		(−1.116)	
	b	−0.014	−0.0195	−0.0020	−0.002
		(−0.62)	(−0.787)	(−0.754)	(−0.589)
Software Development/Services	a	−0.0003	.	0.0011	.
		(−0.048)		(1.658)	
	b	−0.002	−0.0059	−0.0034**	−0.0033*
		(−0.161)	(−0.388)	(−2.115)	(−1.733)
Research, Development and Testing Services	b	−0.084	−0.1265*	−0.0007	−0.0012
		(−1.177)	(−1.936)	(−0.208)	(−0.332)
VARIETY-BASED INDUSTRIES					
Chemicals	a	−0.023	0.0243	0.0007	0.0046
		(−0.773)	(0.841)	(0.169)	(1.136)
	b	−0.0498*	−0.0259	−0.0034	−0.0015
		(−1.93)	(−0.902)	(−0.874)	(−0.357)
Pharmaceuticals and Biological Products	b	−0.342	.	−0.0005	.
		(−1.522)		(−0.191)	
Material Handling Equipment	b	−2.100	.	−0.0133	.
		(−1.229)		(−0.779)	
Other High-tech Manufacturing Industries	a	−0.062**	−0.0543	0.0044	0.004
		(−2.221)	(0.0763)	(0.751)	(0.504)
	b	−0.006	−0.0085	−0.0012	−0.0011
		(−0.89)	(−1.154)	(−0.862)	(−0.598)

NOTE: * = significance at the .10 level, ** = significance at the .05 level (t statistics are in parentheses).

4.6.1 The Growth Patterns of Texan High-tech Industry

With reference to employed civilian labor force, the share of Texas high-tech industry increased rapidly in the period of 1988–1995. As shown in Figure 4.8, the percentage share of high-tech industry increased from 3.82 % in 1988 to 7.16 % in 1995. Within the past decade, Texas has become the home of many high-tech industries. Although an increase of percentage share is by no means synonymous with high-tech job creation, the conclusion is that most high-tech industries are rapidly growing or at least net job generators, although some industries such as military-related industries are not. In particular, as shown in Figure 4.9, the growth rates differ by high-tech sector. Intellectual capital industries and variety-based industries grew rapidly, whereas volatile industries such as aircraft/aerospace and military-related industries declined. Variety-based industries still played an important role in high-tech industry for this period.

On the other hand, Figure 4.10 shows the growth patterns by firm size and organizational type. The growth rate defined by the ratio of employment gains for its own initial level for the period of 1988 to 1995 differs by firm size and organizational type. In terms of employment growth rate, small and single-location firms grew more rapidly than large and multi-locational establishments.

This finding is consistent with recent studies regarding small vs. large firms. For a decade, initiated by Birch's work (1979), small firms have been seen as a major source of job creation. Thus, small firms have become the focus of public policy. The claims concerning the role of small firms are frequently presented as justification for tax incentives, regulatory policies and other government programs. Recently, however, there has been a dispute about the significance of small firms to the rest of the economy. Haltiwanger et al. (1993: 3) found that large plants and firms account for mostly newly-created manufacturing jobs. Smaller manufacturing firms and plants exhibit higher gross rates of job creation but not higher net rates. Survival rates for new and existing manufacturing jobs increase sharply with employee size (Haltiwanger et al., 1993: 1). However, small firms actually tend to create a larger proportion of new jobs than their share of employment in the economy (Erdevig, 1986: 22), although the role of small firms in job creation varies with the period of analysis, the industrial sectors examined, and the geographic locations.

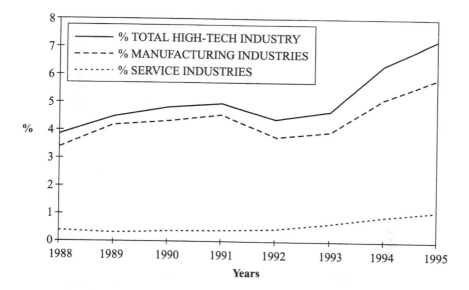

Figure 4.8 Percentage Share of High-tech Industry for Employed Civilian Labor Force, 1988–1995

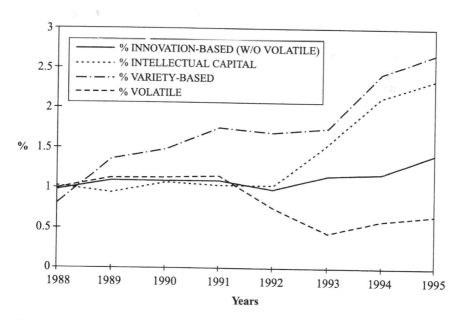

Figure 4.9 Percentage Share of High-tech Industry for Employed Civilian Labor Force by Sector, 1988–1995

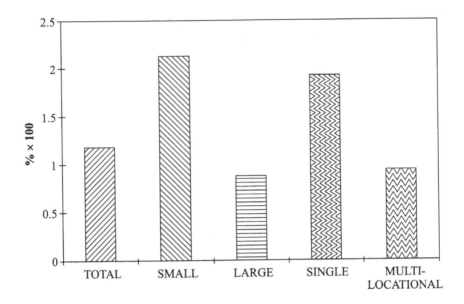

**Figure 4.10 Growth Rates of High-tech Industry by Firm Size and
Organizational Type, 1988–1995**

*4.6.2 Impact of Initial High-tech Employment on Regional Growth of Employed
Civilian Labor Force*

What about the potential of high-tech industry to create new jobs for this period?
First, the results of correlation analysis (Table 4.31) show that there are strong,
positive relationships between employed civilian labor force (CLF) growth between
1990–1995 and high-tech employment growth between 1990 and 1995, and the
initial level of high-tech employment in 1990. The correlation coefficients in
metropolitan areas are much higher than those in municipalities.

Scatter diagrams using cross-sectional data for 1990–1995 show a possible
curvilinear relationship between log employed CLF and initial high-tech level.
Figure 4.11 reveals the relationship between initial high-tech employment level
(1990) and log employed civilian labor force growth (1990–1995). Interestingly,
with reference to the initial high-tech employment level, locations with less than
2000–3000 high-tech employment did not show any positive relationship between
log employed CLF growth and the initial high-tech level. This implies that there is
high-tech industrial churning around this level. Locations with over 2000–3000
employees at the initial level display a positive relationship between these variables.
As the growth rates for selectively chosen locations gradually decrease, the positive
relationship weakens.

Table 4.31 Correlation Coefficients between Initial High-tech Level (1990), High-tech Employment Growth (1990–1995), and Employed Civilian Labor Force Growth (1990–1995)

[MAs & COUNTIES]

	ECLF950	TYG950	TYGSI950	TYGM950	Y90	YSI90	YM90
ECLF950	1						
TYG950	0.92692	1					
TYGSI950	0.92747	0.99356	1				
TYGM950	0.93009	0.98652	0.97152	1			
Y90	0.91057	0.89941	0.92013	0.91361	1		
YSI90	0.88593	0.83587	0.8695	0.84468	0.98391	1	
YM90	0.89744	0.88946	0.90902	0.90577	0.99871	0.98149	1

[MUNICIPALITIES]

	ECLF950	TYG950	TYGSI950	TYGM950	Y90	YSI90	YM90
ECLF950	1						
TYG950	0.71013	1					
TYGSI950	0.74463	0.85853	1				
TYGM950	0.5692	0.91942	0.60994	1			
YT90	0.74258	0.66137	0.79715	0.41176	1		
YSI90	0.68207	0.85502	0.70039	0.74004	0.77997	1	
YM90	0.65589	0.54405	0.7032	0.28428	0.98547	0.69871	1

[MAs & COUNTIES]

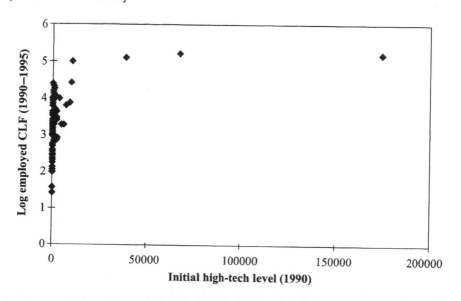

[MUNICIPALITIES]

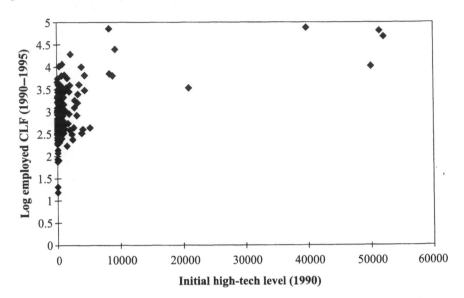

**Figure 4.11 Relationship between Initial High-tech Employment Level (1990)
and Log Employed Civilian Labor Force Growth (1990–1995)**

Regression analysis suggests that the curvilinear relationship is second-degree.

The regression results are laid out in Table 4.32. In model I and model II, equivalent to equation (32), the employment change of employed CLF in logs (LNECLF950) is assumed to be a function of initial high-tech employment level in logs (Y90) and a quadratic term (YSQ), controlling for the percentage of manufacturing industries in 1990 (MANU) and/or the regional dummies (CMSA and MSA). In model III, equivalent to equation (33), the employment change of employed CLF in logs (LNECLF950) is assumed to be a function of the inverse term of the initial high-tech employment level in logs (YRV), controlling for the percentage of manufacturing industries in 1990 (MANU) and/or the regional dummies (CMSA and MSA).

The second-degree model fitted: the quadratic term was negatively significant. This was mainly due to the employment growth of multi-locational establishments (discussed later). The percentage of manufacturing industries in 1990 (MANU) was negatively associated with the employment change of employed CLF in logs (LNECLF950). The higher the initial level of manufacturing industries, the lower the growth of employed CLF. This was especially true when municipalities were used as the units of observations. A higher level of manufacturing industries tends to deter high-tech development, other things being equal.

4.6.3 Differences in the Effects of Initial High-tech Employment on Regional Growth of Employed Civilian Labor Force by Organizational Type

The next question is what effects single vs. multi-locational establishments had on potential job creation. Using cross-sectional data for 1990–1995, the scatter diagram between log employed CLF growth and initial high-tech level is shown in Figure 4.12. It reveals a similar pattern for each group to that of the previous section.

This question is tested by equations (34), (35), and (36). The regression results are shown in Table 4.33. In model I, equivalent to equation (34), the employment change of employed CLF in logs (LNECLF950) is assumed to be a function of initial high-tech employment level of single and multi-locational establishments in logs (YSI90, and YM90, respectively) and the quadratic term for total initial level (YSQ), controlling for the percentage of manufacturing industries in 1990 (MANU) and the regional dummies (CMSA and MSA). In model II and model III, equivalent to equations (35) and (36), respectively, the employment change of employed CLF in logs (LNECLF950) is assumed to be a function of the inverse term of its initial high-tech employment level in logs (YSIRV and YMRV, respectively), controlling for the percentage of manufacturing industries in 1990 (MANU) and the regional dummies (CMSA and MSA).

Table 4.32 Regression Results on the Effects of Initial High-tech Employment on Regional Growth of Employed Civilian Labor Force

DEP. VARIABLE: LOG EMPLOYED CLF (LNECLF950)

	MAs & COUNTIES			MUNICIPALITIES		
VAR.	MODEL I	MODEL II	MODEL III	MODEL I	MODEL II	MODEL III
INTERCEP	2.9541**	3.318427**	3.487353**	2.624798**	3.413241*	3.630659**
	(−18.811)	(−13.677)	(10.152)	(−11.664)	(11.263)	(9.129)
Y90	0.038671**	0.041782**		0.078795**	0.088908**	
	(3.335)	(−3.741)		(3.59)	(4.357)	
YSQ	−0.000148**	−0.00016**		−0.000993**	−0.00122**	
	(−2.739)	(−3.022)		(−2.348)	(−3.082)	
YRV			−15.360428			−11.54763
			(−1.263)			(−1.363)
MANU		−0.021059*	−0.017572		−0.034296**	−0.034838**
		(−1.908)	(−1.326)		(−3.565)	(−2.791)
CMSA	0.109424	−0.224477	4.911**	0.523192**	0.336339	0.441506
	(0.136)	(−0.286)	(4.548)	(2.249)	(1.525)	(1.518)
MSA	0.834094**	0.693903**	2.584**	0.737535**	0.4093*	0.41887
	(4.561)	(3.669)	(2.54)	(3.143)	(1.743)	(1.355)
F-VALUE	16.423	15.2	11.154	13.701	15.477	4.121
R-SQ	0.7164	0.7525	0.6318	0.4574	0.5473	0.2023

ADJ. R-SQ	0.6728	0.703	0.5752	0.4241	0.512	0.1532
N	31	31	31	70	70	70
CON. INDEX	11.9901	12.8508	8.3871	11.9983	14.1492	13.4966

NOTE 1: * = significance at the .10 level, ** = significance at the .05 level.
NOTE 2: The coefficients for Y90, and YSQ are multiplied by 1,000 persons.

[MAs & COUNTIES]

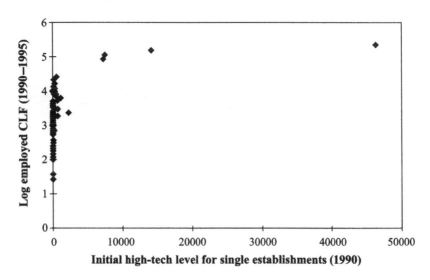

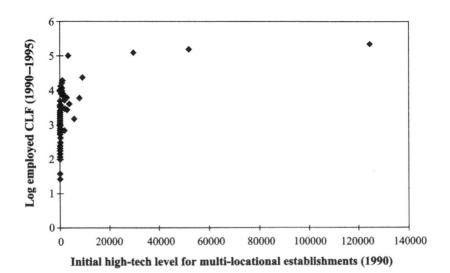

[MUNICIPALITIES]

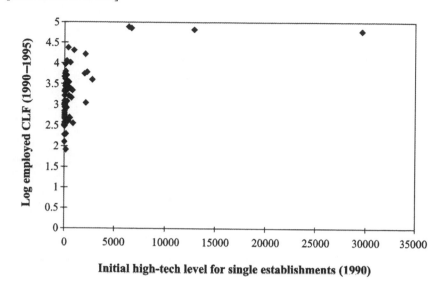

Initial high-tech level for single establishments (1990)

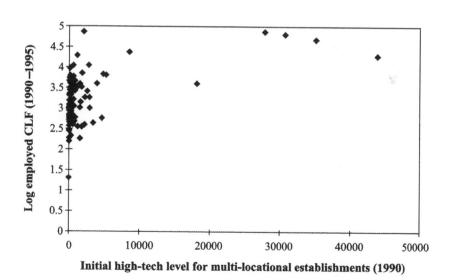

Initial high-tech level for multi-locational establishments (1990)

**Figure 4.12 Relationship between Initial Employment Level for Single and
Multi-locational Establishments (1990) and Log Employed
Civilian Labor Force Growth (1990–1995)**

Table 4.33 Differences in the Effects of Initial High-tech Employment on Regional Growth of Employed Civilian Labor Force by Organizational Type

DEP. VARIABLE: LOG EMPLOYED CLF (LNECLF950)

VAR.	MAs & COUNTIES			MUNICIPALITIES		
	MODEL I	MODEL II	MODEL III	MODEL I	MODEL II	MODEL III
INTERCEP	3.196171**	3.556628**	3.266281**	3.454991**	3.565156**	3.701154**
	(14.408)	(10.791)	(11.045)	(10.57)	(9.064)	(9.161)
YSI90	0.143**			0.045012**		
	(2.896)			(2.078)		
YM90	0.03099*			0.067496**		
	(1.876)			(2.939)		
YSQ	−0.000271**			−0.000686*		
	(−4.369)			(−1.839)		
YSIRV		−2.440996*			−2.337755*	
		(−1.767)			(−1.883)	
YMRV			−2.081524			−2.695084
			(−0.655)			(−1.29)
MANU	−0.016241	−0.01821	−0.013249	−0.033127**	−0.032112**	−0.037524**
	(−1.621)	(−1.418)	(−0.992)	(−3.159)	(−2.594)	(−2.946)
CMSA	0.014114	1.956733**	2.181499**	0.332559	0.501565*	0.412743
	(0.021)	(4.914)	(5.558)	(1.411)	(1.726)	(1.422)
MSA	0.752829**	0.644885**	0.869219**	0.428832*	0.470716	0.395777
	(4.444)	(2.608)	(3.868)	(1.704)	(1.535)	(1.277)
F-VALUE	16.927	12.132	10.409	10.279	4.635	4.063
R-SQ	0.8089	0.6511	0.6156	0.4947	0.2219	0.2
ADJ. R-SQ	0.7611	0.5975	0.5564	0.4465	0.1741	0.1508
N	31	31	31	70	70	70
CON. INDEX	17.7831	8.3198	7.2702	15.096	13.8825	13.6498

NOTE 1: * = significance at the .10 level, ** = significance at the .05 level.
NOTE 2: The coefficients for YSI90, YM90, and YSQ are multiplied by 1,000 persons.

The regression analysis suggests that the relationship is curvilinear in the second-degree for single establishments. Both the coefficients of the quadratic term (YSQ) and the inverse term (YSIRV) were negatively significant. On the other hand, for multi-locational establishments, the relationship between the initial employment level of multi-locational establishments in logs and the employment change of employed CLF in logs (LNECLF950) fitted the second-degree curve. Only the quadratic term was negatively significant.

Thus, in terms of job creation, the initial level of multi-locational establishments also is positively associated with employed CLF. The effects of the initial level of multi-locational establishments on employed CLF seem to be more readily observable. This is one of the justifications for attracting multi-locational establishments. However, another important point is that any specialization is vulnerable to unpredictable and perhaps abrupt changes in demand for a location's exports (Malecki, 1991: 55). In effect, multi-locational establishments were strongly attracted to specialized locations. The employment growth of multi-locational establishments is a major factor responsible for specialization.

Chapter 5

Summary and Policy Implications

The major objective of this study was to explore the regional high-tech employment effects of dynamic externalities, and the differences by firm size, organizational type, and product. Three arguable hypotheses suggested by the literature reviewed were tested:

1. Regional growth in high-tech employment is determined by endogenous technological progress.
2. The growth of small, single-location firms is more likely to be driven by endogenous processes than is the growth of large, multi-locational firms.
3. Endogenous growth is more likely to occur in competitive local economies rather than in local economies dominated by one or relatively few large firms.

The hypotheses were tested using the data for the state of Texas for the years 1988–1995.

Under the assumption that there is a bi-directional feedback relationship between regional high-tech employment growth and dynamic externalities, I formulated a simultaneous equation model. Cross-sectional and time-series analyses were employed to test the three hypotheses. In addition to OLS estimation, the two-stage least squares estimation and a Hausman specification test were employed for regression analyses. For estimating the models, I used two different units of analysis: metropolitan areas and non-metropolitan counties, and municipalities. The following are the major empirical findings:

First, the results strongly support hypothesis 1. Past specialization significantly affected regional high-tech employment growth, and in turn higher initial high-tech employment level and growth led to more specialized locations. The MAR effects were quantitatively large. As Henderson (1994) argues, specialization had about 5–6 years of time-lag in metropolitan areas. In addition, a very high degree of persistence was observed across metropolitan areas and municipalities, despite strong evidence for relocation.

On the other hand, Jacobs effects were not readily observable for high-tech industry as a whole. However, when the data were controlled for firm size and high-tech product, they were partially observed for small firms and such products as aerospace/aircraft, measurement and testing equipment, software, and other manufacturing industries. Innovation-based industries and intellectual-capital industries tended to have relatively strong diversity effects. However, time-series analysis revealed that the persistence of diversity effects was not strong.

Second, with respect to small vs. large firms, the results partially support hypothesis 2. Hypothesis 2 was true for Jacobs effects, but not for MAR effects.

Small firms had larger Jacobs effects than large firms. On the other hand, large firms had larger MAR effects than small firms. These findings match recent trends. Although economies of scale can be realized by ever-smaller businesses, more diversified environments are necessary for small firms lacking internal knowledge networks to grow. High-tech enclaves can help small firms achieve diversified environments.

With respect to single vs. multi-locational establishments, hypothesis 2 is rejected. Multi-locational establishments had almost equal or a little stronger MAR effects than single firms. This implies that multi-locational establishments responded quickly to dynamic externalities. Jacobs effects were not observed for either organizational type.

Finally, the test results for individual high-tech product support hypothesis 3. For some products such as medical equipment/devices and material handling equipment, local competition had significantly positive effects on their own employment growth, whereas for other products such as measurement and testing equipment and chemicals, local competition had significantly negative effects. One possible interpretation is that local competition affects employment growth differentially, depending on the cooperational structure within industries. This supports the Porter's and Jacobs' views, and also is consistent with the findings of Glaeser et al. (1991). Local competition also had indirect effects on employment growth via dynamic externalities. Whereas local competition deterred current specialization, it facilitated significantly diversity for such products as software and military industries.

In addition to testing the three hypotheses, I attempted to examine the effects of high-tech development on regional economic growth, and the differences by organizational type. I hypothesized that initial high-tech employment level is positively associated with employed CLF growth. To describe the trend correctly, nonlinear terms were used. The models were tested by ordinary least squares estimation method, using cross-sectional data for the years 1990–1995. The findings are as follows:

First, the scatter diagram between log employed CLF growth and initial high-tech level showed that beyond a threshold of 2000–3000 employees there was a positive relationship between these variables. The positive relationship weakened as this high-tech employment threshold increased. This pattern was much more apparent for multi-locational establishments than for single firms.

Second, regression analyses showed that the relationship between log employed CLF and initial high-tech level was nonlinear in the second-degree both for single firms and for multi-locational establishments.

These empirical findings strongly support endogenous growth theories. There are several important policy implications. First, unlike neo-classical theory, endogenous models predict long-term divergence in regional high-tech growth and increasing returns to scale. If growth is endogenous, effective economic development policy can be produced by local initiatives. For large firms, local initiatives can support local infrastructure programs to maximize scale effects. For small firms, local initiatives can help improve industrial conditions for new entry. Such development programs as entrepreneur education projects and informational network systems are important examples. Since initial conditions matter, a contingent approach that considers the local context is more appropriate than 'one-style fits all' federal policy-making. Regional government should have real authority and control over its own economic development policy-making.

Second, attraction of small and large firms represents complementary rather than competing policy alternatives. Large firms account for mostly newly-created manufacturing jobs, but large firms are more sensitive to recession and recessionary periods. Small firms tend to create a larger proportion of new jobs than their share of employment in the economy. Even in weak or declining regions, small firms are important sources of employment growth. Importantly, small firms are most successful if located in diversified environments. Diversification is essential if metropolitan areas are to develop the capacity to adapt to economic and technological changes.

Third, the choice of single vs. multi-locational establishments also represents complementary rather than competing policy alternatives. Attracting multi-locational establishments provides immediate rewards in terms of jobs and income. Multi-locational establishments are strongly attracted to specialized locations. In turn, the employment growth of multi-locational establishments produces greater specialization. On the other hand, encouraging regional business initiative tends to lessen dependence on externally controlled firms, facilitating local adaptation to the continuing process of technical change.

Fourth, local competition can enhance the effects of knowledge spillovers directly or indirectly via dynamic externalities. The effects of local competition seem to depend on the cooperational structure within industries.

Finally, high-tech industry remains a powerful source of job creation. With reference to employed civilian labor force, the share of Texas high-tech industry increased rapidly from 3.82 % in 1988 to 7.16 % in 1995. Most high-tech industries are rapidly growing or at least net job generators, although some industries (e.g., military-related industries) are not. The unavoidable conclusion is that technological externalities associated with knowledge spillovers can be one of the important 'recipes' for promoting high-tech development.

In this study, I dealt with theoretically arguable hypotheses regarding the nature of high-tech development. The evidence produced has important implications for Texas industrial policy. However, the study focuses upon a particular recent period in the history of Texas in which high-tech industry has grown rapidly. As a result, the finding may not be applicable for other time periods or places – more empirical work is needed. As for policy implications, I only dealt with the differences by firm size, organizational type, and product. Future studies should address the broader set of conditions under which dynamic externalities are maximized.

Methodologically, some limitations of this study should be noted. In my modeling, I assumed that only high-tech employment growth and externalities are currently and simultaneously determined. With longer time-series data, more dynamic modeling could be developed to capture the linkage effects with variables such as wages. Policy measures also might be inserted as endogenous variables. I used regional dummies to capture other local variables, but there still could be some omission bias. Further investigation is needed of the specific local factors that affect the ability to generate endogenous growth. With such investigation, it appears clear that the alternative theoretical framework validated in this investigation will provide new and important support for economic development activities.

APPENDICES

Appendix A

The Estimation Model of the Regional High-tech Employment Effects of Dynamic Externalities

Glaeser et al. (1991) and Henderson et al. (1995) developed methods for estimating the regional effects of dynamic externalities cross-sectionally. In this study I modified Henderson et al.'s method. This is reflected in equation (1) of the basic model in page 23. As in Henderson et al.'s method, regional high-tech employment growth is assumed to be a function of historical specialization and diversity, and the initial high-tech employment level. This equation builds on the basis that the equilibrium employment level is a reduced form for the following equations:

For a high-tech product k, in location i at time t, the equilibrium employment level is such that the local wage rate (Wit) equals the value of marginal product (VMP), or

$Wit = Ait\, f'(Yit; \ldots)$
where Ait = the state of technology for high-tech product k,
$Ait\, f(Yit; \ldots)$ = industry output for high-tech product k,
Yit = current employment level of high-tech product k.

The state of technology is postulated as a function of historical specialization and diversity, and the initial employment level of high-tech product k, or

$$Ait = g(\sum_{m=0}^{6} SPk(i,t-m), \sum_{m=0}^{6} Dk(i,t\ m), Yk(i1))$$

where $\sum_{m=0}^{6} SPk(i,t-m)$ = historical specialization of product k with time-lag $m=0\ldots6$,

$\sum_{m=0}^{6} Dk(i,t-m)$ = historical diversity of product k with time-lag $m=0\ldots6$,

$Yk(i1)$ = initial employment level of high-tech product k.

In addition, the price of the high-tech product k (Pit) is given by the inverse demand function of high-tech employment and regional characteristics, or

$Pit = h(Yit, RCit)$
where $RCit$ = regional characteristics.

The List of Municipalities with High-tech Firms as of 1995 and Relevant Metropolitan Areas and Non-metropolitan Counties

No.	Municipalities	Relevant Metropolitan Areas and Non-metropolitan Counties
1	Abilene	Abilene MSA
2	Addison	Dallas-FW CMSA
3	Alamo	McAllen-Edinberg-Mission MSA
4	Albany	Schackelford
5	Alice	Jim Wells
6	Allen	Dallas-FW CMSA
7	Alvarado	Dallas-FW CMSA
8	Alvin	Houston-Galveston-Brazoria CMSA
9	Amarillo	Amarillo MSA
10	Angleton	Houston-Galveston-Brazoria CMSA
11	Annona	Red River
12	Arcadia	Unknown
13	Argyle	Dallas-FW CMSA
14	Arlington	Dallas-FW CMSA
15	Atascosa	Unknown
16	Athens	Henderson
17	Atlanta	Cass
18	Aubrey	Dallas-FW CMSA
19	Austin	Austin-San Marcos MSA
20	Avalon	Dallas-FW CMSA
21	Azle	Dallas-FW CMSA
22	Balch Springs	Dallas-FW CMSA
23	Ballinger	Runnels
24	Bandera	Bandera
25	Bastrop	Bastrop
26	Bay City	Matagorda
27	Baytown	Houston-Galveston-Brazoria CMSA
28	Beaumont	Beaumont-Port Arthur MSA
29	Bedford	Dallas-FW CMSA
30	Beeville	Bee
31	Bellaire	Houston-Galveston-Brazoria CMSA

32	Bellville	Austin
33	Belton	Killeen-Temple MSA
34	Big Spring	Howard
35	Bishop	Corpus Christi MSA
36	Boerne	Kendall
37	Bonham	Fannin
38	Borger	Hutchinson
39	Bowie	Montague
40	Boyd	Wise
41	Breckenridge	Stephens
42	Brenham	Washington
43	Brookshire	Houston-Galveston-Brazoria CMSA
44	Brownsville	Brownsville-Harlingen-San Benito MSA
45	Brownwood	Brown
46	Bruni	Laredo MSA
47	Bryan	Bryan-College Station MSA
48	Bryson	Jack
49	Buda	Austin-San Marcos MSA
50	Burleson	Dallas-FW CMSA
51	Burnet	Burnet
52	Canutillo	El Paso MSA
53	Canyon	Amarillo MSA
54	Canyon Lake	San Antonio MSA
55	Carrollton	Dallas-FW CMSA
56	Cedar Hill	Dallas-FW CMSA
57	Cedar Park	Austin-San Marcos MSA
58	Center Point	Kerr
59	Centerville	Leon
60	Channelview	Houston-Galveston-Brazoria CMSA
61	Cibolo	San Antonio MSA
62	Cisco	Eastland
63	Cleburne	Dallas-FW CMSA
64	Cleveland	Houston-Galveston-Brazoria CMSA
65	Clifton	Bosque
66	Clute	Houston-Galveston-Brazoria CMSA
67	Clyde	Callahan
68	College Station	Bryan-College Station MSA
69	Colleyville	Dallas-FW CMSA
70	Commerce	Hunt
71	Conroe	Houston-Galveston-Brazoria CMSA
72	Coppell	Dallas-FW CMSA
73	Corpus Christi	Corpus Christi MSA
74	Corsicana	Navarro
75	Crockett	Houston
76	Crosby	Houston-Galveston-Brazoria CMSA
77	Crowley	Dallas-FW CMSA
78	Cuero	DeWitt

79	Cypress	Houston-Galveston-Brazoria CMSA
80	Dallas	Dallas-FW CMSA
81	Dayton	Houston-Galveston-Brazoria CMSA
82	Decatur	Wise
83	Deer Park	Houston-Galveston-Brazoria CMSA
84	Del Rio	Val Verde
85	Denison	Sherman-Denison MSA
86	Denton	Dallas-FW CMSA
87	Desoto	Dallas-FW CMSA
88	Diboll	Angelina
89	Dickinson	Houston-Galveston-Brazoria CMSA
90	Dike	Hopkins
91	Douglassville	Cass
92	Dripping Springs	Austin-San Marcos MSA
93	Duncanville	Dallas-FW CMSA
94	Eagle Pass	Maverick
95	East Bernard	Wharton
96	Ector	Fannin
97	Edgewood	VanZandt
98	Edinberg	McAllen-Edinberg-Mission MSA
99	Edna	Jackson
100	El Paso	El Paso MSA
101	Elgin	Bastrop
102	Elmendorf	San Antonio MSA
103	Ennis	Dallas-FW CMSA
104	Euless	Dallas-FW CMSA
105	Eustace	Henderson
106	Everman	Dallas-FW CMSA
107	Fairview	Dallas-FW CMSA
108	Farmers Branch	Dallas-FW CMSA
109	Farmersville	Dallas-FW CMSA
110	Flatonia	Gonzales
111	Floresville	Wilson
112	Flower Mound	Dallas-FW CMSA
113	Forney	Dallas-FW CMSA
114	Fort Worth	Dallas-FW CMSA
115	Fredericksburg	Gillespie
116	Freeport	Houston-Galveston-Brazoria CMSA
117	Friendswood	Houston-Galveston-Brazoria CMSA
118	Frisco	Dallas-FW CMSA
119	Gainesville	Cooke
120	Galena Park	Houston-Galveston-Brazoria CMSA
121	Galveston	Houston-Galveston-Brazoria CMSA
122	Garden City	Glasscock
123	Garland	Dallas-FW CMSA
124	Gatesville	Killeen-Temple MSA
125	Georgetown	Austin-San Marcos MSA

126	Giddings	Lee
127	Gilmer	Upshur
128	Gladewater	Longview-Marshall MSA
129	Godley	Dallas-FW CMSA
130	Graham	Young
131	Granbury	Hood
132	Grand Prairie	Dallas-FW CMSA
133	Grapevine	Dallas-FW CMSA
134	Greenville	Hunt
135	Groves	Beaumont-Port Arthur MSA
136	Haltom City	Dallas-FW CMSA
137	Harlingen	Brownsville-Harlingen-San Benito MSA
138	Hawkins	Wood
139	Hempstead	Houston-Galveston-Brazoria CMSA
140	Hereford	Deaf Smith
141	Hewitt	Waco MSA
142	Highlands	Houston-Galveston-Brazoria CMSA
143	Hobson	Karnes
144	Hondo	Medina
145	Houston	Houston-Galveston-Brazoria CMSA
146	Howe	Sherman-Denison MSA
147	Humble	Houston-Galveston-Brazoria CMSA
148	Huntsville	Walker
149	Hurst	Dallas-FW CMSA
150	Hutchins	Dallas-FW CMSA
151	Hutto	Austin-San Marcos MSA
152	Ingleside	Corpus Christi MSA
153	Ingram	Kerr
154	Irving	Dallas-FW CMSA
155	Jacksonville	Cherokee
156	Jasper	Jasper
157	Jewett	Leon
158	Johnson City	Blanco
159	Joshua	Dallas-FW CMSA
160	Judson	Unknown
161	Karnes City	Karnes
162	Katy	Houston-Galveston-Brazoria CMSA
163	Kaufman	Dallas-FW CMSA
164	Keller	Dallas-FW CMSA
165	Kenedy	Karnes
166	Kennedale	Dallas-FW CMSA
167	Kerrville	Kerr
168	Kilgore	Longview-Marshall MSA
169	Kingsville	Kleberg
170	Kingwood	Houston-Galveston-Brazoria CMSA
171	Kountze	Beaumont-Port Arthur MSA
172	La Feria	Brownsville-Harlingen-San Benito MSA

173	La Grange	Fayette
174	La Marque	Houston-Galveston-Brazoria CMSA
175	La Porte	Houston-Galveston-Brazoria CMSA
176	Lago Vista	Austin-San Marcos MSA
177	Lake Dallas	Dallas-FW CMSA
178	Lake Jackson	Houston-Galveston-Brazoria CMSA
179	Lakehills	Bandera
180	Lamesa	Dawson
181	Lancaster	Dallas-FW CMSA
182	Laredo	Laredo MSA
183	Latexo	Houston
184	League City	Houston-Galveston-Brazoria CMSA
185	Leander	Austin-San Marcos MSA
186	Levelland	Hockley
187	Lewisville	Dallas-FW CMSA
188	Linden	Cass
189	Livingston	Polk
190	Lockhart	Caldwell
191	Longmott	Calhoun
192	Longview	Longview-Marshall MSA
193	Lubbock	Lubbock MSA
194	Lufkin	Angelina
195	Lumberton	Beaumont-Port Arthur MSA
196	Magnolia	Houston-Galveston-Brazoria CMSA
197	Malakoff	Henderson
198	Manchaca	Unknown
199	Manor	Austin-San Marcos MSA
200	Mansfield	Dallas-FW CMSA
201	Manvel	Houston-Galveston-Brazoria CMSA
202	Marble Falls	Burnet
203	Marshall	Longview-Marshall MSA
204	Masterson	Moore
205	McAllen	McAllen-Edinberg-Mission MSA
206	Mckinney	Dallas-FW CMSA
207	Meridian	Bosque
208	Mesquite	Dallas-FW CMSA
209	Mexia	Limestone
210	Midland	Midland MSA
211	Midlothian	Dallas-FW CMSA
212	Millsap	Dallas-FW CMSA
213	Mineola	Wood
214	Mineral Wells	Palo Pinto, Parker
215	Mission	McAllen-Edinberg-Mission MSA
216	Missouri City	Houston-Galveston-Brazoria CMSA
217	Mont Belvieu	Chambers
218	Montgomery	Houston-Galveston-Brazoria CMSA
219	Morton	Cochran

220	Moulton	Lavaca
221	Mount Pleasant	Titus
222	Nacogdoches	Nacogdoches
223	Nash	Texarkana-T. MSA
224	Navasota	Grimes
225	Nederland	Beaumont-Port Arthur MSA
226	New Braunfels	San Antonio MSA
227	Newark	Wise
228	Odessa	Odessa MSA
229	Old Ocean	Houston-Galveston-Brazoria CMSA
230	Orange	Beaumont-Port Arthur MSA
231	Pampa	Gray
232	Paris	Lamar
233	Pasadena	Houston-Galveston-Brazoria CMSA
234	Pearland	Houston-Galveston-Brazoria CMSA
235	Pflugerville	Austin-San Marcos MSA
236	Pharr	McAllen-Edinberg-Mission MSA
237	Plainview	Hale
238	Plano	Dallas-FW CMSA
239	Point Comfort	Calhoun
240	Port Arthur	Beaumont-Port Arthur MSA
241	Port Isabel	Brownsville-Harlingen-San Benito MSA
242	Port Lavaca	Calhoun
243	Port Neches	Beaumont-Port Arthur MSA
244	Porter	Houston-Galveston-Brazoria CMSA
245	Princeton	Dallas-FW CMSA
246	Prosper	Dallas-FW CMSA
247	Queen City	Cass
248	Quinlan	Hunt
249	Quitman	Wood
250	Refugio	Refugio
251	Richardson	Dallas-FW CMSA
252	Richmond	Houston-Galveston-Brazoria CMSA
253	Roanoke	Dallas-FW CMSA
254	Robstown	Corpus Christi MSA
255	Rockport	Aransas
256	Rockwall	Dallas-FW CMSA
257	Rosenberg	Houston-Galveston-Brazoria CMSA
258	Rosharon	Houston-Galveston-Brazoria CMSA
259	Round Rock	Austin-San Marcos MSA
260	Rowlett	Dallas-FW CMSA
261	Royse City	Dallas-FW CMSA
262	Sachse	Dallas-FW CMSA
263	Saginaw	Dallas-FW CMSA
264	San Angelo	San Angelo MSA
265	San Antonio	San Antonio MSA
266	San Benito	Brownsville-Harlingen-San Benito MSA

267	San Marcos	Austin-San Marcos MSA
268	San Saba	San Saba
269	Sanger	Dallas-FW CMSA
270	Santa Fe	Houston-Galveston-Brazoria CMSA
271	Santa Rosa	Brownsville-Harlingen-San Benito MSA
272	Schertz	San Antonio MSA
273	Scottsville	Longview-Marshall MSA
274	Seabrook	Houston-Galveston-Brazoria CMSA
275	Seagraves	Gaines
276	Sealy	Austin
277	Seguin	San Antonio MSA
278	Selma	San Antonio MSA
279	Seminole	Gaines
280	Sherman	Sherman-Denison MSA
281	Silsbee	Beaumont-Port Arthur MSA
282	Skellytown	Carson
283	Slaton	Lubbock MSA
284	Smithville	Bastrop
285	Snyder	Scurry
286	South Houston	Houston-Galveston-Brazoria CMSA
287	Southlake	Dallas-FW CMSA
288	Splendora	Houston-Galveston-Brazoria CMSA
289	Spring	Houston-Galveston-Brazoria CMSA
290	Stafford	Houston-Galveston-Brazoria CMSA
291	Stanton	Martin
292	Stephenville	Erath
293	Sugar Land	Houston-Galveston-Brazoria CMSA
294	Sulphur Springs	Hopkins
295	Sundown	Hockley
296	Sunnyvale	Dallas-FW CMSA
297	Sunray	Moore
298	Sweeny	Houston-Galveston-Brazoria CMSA
299	Sweetwater	Nolan
300	Taylor	Austin-San Marcos MSA
301	Temple	Killeen-Temple MSA
302	Terrell	Dallas-FW CMSA
303	, Texarkana	Texarkana-T. MSA
304	Texas City	Houston-Galveston-Brazoria CMSA
305	The Colony	Dallas-FW CMSA
306	The Woodlands	Houston-Galveston-Brazoria CMSA
307	Tilden	McMullen
308	Tomball	Houston-Galveston-Brazoria CMSA
309	Tyler	Tyler MSA
310	Universal City	San Antonio MSA
311	Uvalde	Uvalde
312	Vernon	Wilbarger
313	Victoria	Victoria MSA

314	Waco	Waco MSA
315	Waller	Houston-Galveston-Brazoria CMSA
316	Warda	Fayette
317	Waskom	Longview-Marshall MSA
318	Waxahachie	Dallas-FW CMSA
319	Weatherford	Dallas-FW CMSA
320	Webster	Houston-Galveston-Brazoria CMSA
321	Weimar	Colorado
322	Weslaco	McAllen-Edinberg-Mission MSA
323	Westlake	Dallas-FW CMSA
324	Wetmore	Unknown
325	White Oak	Longview-Marshall MSA
326	White Settlement	Dallas-FW CMSA
327	Whitesboro	Sherman-Denison MSA
328	Whitewright	Sherman-Denison MSA
329	Whitney	Hill
330	Whitsett	Live Oak
331	Wichita Falls	Wichita Falls MSA
332	Willis	Houston-Galveston-Brazoria CMSA
333	Winnsboro	Wood, Franklin
334	Wylie	Dallas-FW CMSA

Appendix C

Hausman Specification Test Procedure

This is a test of whether the 2SLS estimates are significantly different from the OLS estimates. In this study, as a statistical criterion, the Hausman specification test is employed to choose between a simultaneous equation model that permits simultaneity and one that does not.

The basic model that was employed in this study is:

$$YGk(it) = f\left(\sum_{m=0}^{6} SPk(i,t-m), \sum_{m=0}^{6} Dk(i,t-m), Wk(i,t), X(i1), Yk(i1), DM\right) \qquad (1)$$

$$SPk(it) = g\left(TYGk(it), \sum_{m=0}^{6} Wk(i,tm), X(i1), Yk(i1), DM\right) \qquad (2)$$

$$Dk(it) = h\left(TYGk(it), \sum_{m=0}^{6} Wk(i,t-m), X(i1), TYGo(it), Yk(i1), Yo(i1), DM\right) \qquad (3)$$

Suppose we test whether in equation (1) the regressors $SPk(it)$ and $Dk(it)$ were correlated with the disturbance term. Under the null hypothesis that in large samples, the fitted values of $SPk(it)$ and $Dk(it)$ are uncorrelated with the disturbance term, OLS estimation results in consistent and efficient estimates of the parameters, whereas 2SLS yields consistent estimates. On the other hand, under the alternative hypothesis that in large samples, the fitted values of $SPk(it)$ and $Dk(it)$ are correlated with the disturbance term, 2SLS yields consistent estimates, but OLS does not. The procedure for the Hausman specification test employed in this study is as follows:

1. Estimate the reduced form equations (2) and (3) by OLS and retrieve the fitted values from these equations.
2. Estimate by OLS the expanded equation (1) with added regressors, that is, the fitted values of $SPk(it)$ and $Dk(it)$.
3. Test the null hypothesis that the coefficients for the added regressors = 0.
4. In the same manner, in equations (2) and (3) test the correlation of the fitted value of $TYGk(it)$ and the disturbance terms of equations (2) and (3), respectively.

Bibliography

Arrow, K. J., 'The Economic Implication of Learning by Doing,' *The Review of Economic Studies*, 29(1962): 155–173.

Barkley, David L., 'The Decentralization of High-Technology Manufacturing to Nonmetropolitan Areas,' *Growth and Change*, Winter 1988: 13–30.

Berry, Brian J. L., *Long-Wave Rhythms in Economic Development and Political Behavior*, Baltimore: The Johns Hopkins University Press, 1991.

Birch, D. L., *The Job Generation Process*, M. I. T. Program on Neighborhood and Regional Change, Cambridge: Mass., 1979.

Calem, Paul S.; Carlino, Gerald A., 'Urban Agglomeration Economies in the Presence of Technical Change,' *Journal of Urban Economics*, 29(1991): 82–95.

Choi, Kwang, *Theories of Comparative Economic Growth*, Ames: The Iowa State University Press, 1983.

Cyert, R. M.; Mowery, D. C., *Technology and Employment: Innovation and Growth in the U.S. Economy*, Washington, DC: National Academy Press, 1987.

Erdevig, Eleanor H., 'Small Business, Big Job Growth,' *Economic Perspectives*, Federal Reserve Bank of Chicago, 1986: 15–24.

Follain, James R., 'Comments on Cities, Information, and Economic Growth,' *CityScape*, August 1994, 1:1, 49–51.

Glaeser, 'Cities, Information, and Economic Growth,' *Cityscape*, August 1994, 1:1, 9–47.

Glaeser, Edward L.; Kallal, Hedi D.; Scheinkman, Jose A.; Shleifer, Andrei, *Growth in Cities*, Working Paper No. 3787, Cambridge, Mass.: NBER, July 1991.

_____; Maré, David C., *Cities and Skills*, Working Paper No. 4728, Cambridge, Mass.: NBER, May 1994.

Glickman, Norman J., 'Comments on Papers by Henderson, Andrews, and Garcia-Milà and McGuire,' *Cityscape*, August 1994, 1:1, 117–121.

Goldstein, Harvey A.; Luger, Michael I., 'Theory and Practice in High-Tech Economic Development,' in *Theories of Local Economic Development: Perspectives From Across the Disciplines?*, edited by Richard D. Bingham and Robert Mier, Newbury Park: SAGE Publications, Inc., 1993: 147–171.

Hall, Peter G., 'The Geography of High Technology: An Anglo-American Comparison,' in *The Spatial Impact of Technological Change*, edited by John F. Brotchie, Peter Hall, and Peter W. Newton, New York: Croom Helm, 1987.

_____; Preston, P., *The Carrier Wave: New Information Technology and the Geography of Innovation 1846–2003*, London: Unwin Hyman, 1988.

Haltiwanger, John; Davis, Stephen; Schuh, Scott, 'Small Business and Job Creation: Dissecting the Myth and Reassessing the Facts,' National Bureau of Economic Research (NBER) No. 4492, October, 1993.

Harris, Candee S., 'Establishing High-Technology Enterprises in Metropolitan Areas,' in *Local Economies in Transition: Policy Realities and Development Potentials*, edited by Edward M. Bergman, Durham: Duke University Press, 1986: 165–184.

Henderson, Vernon, 'Efficiency of Resource Usage and City Size,' *Journal of Urban Economics*, 19(1986), 47–70.

_____, 'Externalities and Industrial Development,' *Cityscape*, August 1994, 1:1, 75–93.

_____; Kuncoro, Ari; Turner, Matthew, 'Industrial Development in Cities,' *Journal of Political Economy* 103(October 1995): 1067–1090.

Jacobs, Jane, *The Economy of Cities*, New York: Vintage Books, 1968.

_____, *Cities and the Wealth of Nations*, New York: Vintage Books, 1984.

Jaffe, Adam B.; Trajtenberg, Manuel; Henderson, Rebecca, 'Geographic Localization of Knowledge Spillovers as Evidenced by Patent Citations,' *Quarterly Journal of Economics*, 108(1993): 577–598.

Joint Economic Committee, U.S. Congress, *Location of High Technology Firms and Regional Economic Development*, Washington, D.C.: Government Printing Office, June, 1982.

Jones, Charles, 'Time Series Tests of Endogenous Growth Models,' *Quarterly Journal of Economics*, (1995): 495–525.

Kaldor, N., *Further Essays on Economic Theory*, New York: Holmes & Meier, 1978.

Lucas, Robert E., 'On the Mechanics of Economic Development,' *Journal of Monetary Economics*, 22(1988): 3–42.

Lyons, Donald, 'Agglomeration Economies among High Technology Firms in Advanced Production Areas: The case of Denver/Boulder,' *Regional Studies*, 29:3, 265–278.

Malecki, Edward J., 'High-Technology Sectors and Local Economic Development,' in *Local Economies in Transition: Policy Realities and Development Potentials*, edited by Edward M. Bergman, Durham: Duke University Press, 1986: 129–142.

_____, 'What about people in High-Technology? Some Research and Policy Considerations,' *Growth and Change*, Winter 1989, 67–79.

_____, *Technology and Economic Development: the Dynamics of Local, Regional, and National Change*, New York: Longman Scientific & Technical, 1991.

Mansfield, Edwin, *Microeconomics: Theory and Application*, 8th ed., New York: W.W. Norton & Company, Inc., 1994.

Markusen, Ann R., *Profit Cycles, Oligopoly, and Regional Development*, Cambridge: MIT Press, 1985.

_____; Hall, Peter G.; Glasmeier, Amy K., *High Tech America: The What, How, Where, and Why of the Sunrise Industries*, Boston: Allen & Unwin, Inc., 1986.

Marshall, Alfred, *Principles of Economics*, London: Macmillan, 1890.

Miracky, W. F., 'Technological Spillovers, the Product Cycle and Regional Growth,' Manuscript. Cambridge: Massachusetts Inst. Tech., 1992.

Muniak, Dennis C., 'Economic Development, National High Technology Policy and America's Cities,' *Regional Studies*, 1995, 28:8, 803–809.

Office of Technology Assessment, U.S. Congress, *Technology, Innovation and Regional Economic Development*, Washington, D.C.: Government Printing Office, 1984.

Ó hUallacháin, Breandán; Satterthwaite, Mark A., 'Sectoral Growth Patterns at the Metropolitan Level: An Evaluation of Economic Development Incentives,' *Journal of Urban Economics*, 31(1992): 25–78.

Pack, Howard, 'Endogenous Growth Theory: Intellectual Appeal and Empirical Shortcomings,' *Journal of Economic Perspectives*, Winter 1994, 8:1, 55–72.

Petrakos, George C., 'Urban Concentration and Agglomeration Economies: Re-examining the Relationship,' *Urban Studies*, 1992, 29:8, 1219–1229.

Pollard, Jane; Storper, Michael, 'A Tale of Twelve Cities: Metropolitan Employment Change in Dynamic Industries in the 1980s,' *Economic Geography*, January 1996, 72:1, 1–22.

Porter, Michael E., *The Competitive Advantage of Nations*, New York: Free Press, 1990.

Prager, Adam J.; Benowitz, Philip; Schein, Robert, 'Local Economic Development: Trends and Prospects,' *Municipal Year Book 1995*, The International City/County Management Association: Washington, D.C., 1995.

Rauch, James E., 'Productivity Gains from Geographic Concentration of Human Capital: Evidence from the Cities,' *Journal of Urban Economics*, 34(1993), 380–400.

Romer, Paul M., 'Increasing Returns and Long-Run Growth,' *Journal of Political Economy* 94(October 1986): 1002–1037.

_____, 'Growth Based on Increasing Returns Due to Specialization,' *American Economic Review*, 1987, 77:2, 56–62.

_____, 'Human Capital and Growth: Theory and Evidence,' *Carnegie-Rochester Conference Series on Public Policy*, 32(1990): 251–286.

Schumpeter, Joseph A., *Capitalism, Socialism and Democracy*, 3rd ed., New York: Harper and Row, Publishers, Inc., 1950.

Scott, Allen J.; Storper, Michael, 'High Technology Industry and Regional Development: A Theoretical Critique and Reconstruction,' *International Social Science Journal*, 112(May 1987): 215–232.

Shachar, Arie; Felsenstein, Daniel, 'Urban Economic Development and High technology Industry,' *Urban Studies*, 1992, 29:6, 839–855.

Smith, Adam, *An Inquiry Into the Nature and Causes of the Wealth of the Nations*, edited by R. H. Campbell and A. S. Skinner, Indianapolis: Liberty Fund, 1981.

Solow, R. M., 'Contribution to the Theory of Economic Growth,' *Quarterly Journal of Economics*, 70(1956): 65–94.

Texas High Technology Directory, Ashland: Leading Edge Communications, Inc., 1988–1989, 1989–1990, 1990–1991, 1991–1992, 1993, 1994, 1995 & 1996.

The Economist, 'The Puzzling Infirmity of America's Small Firms,' February 18, 1995, 63–64.

Verdoorn, P. J., 'Verdoorn's Law in Retrospect: A Comment,' *Economic Journal*, 90(1980): 382–385.

Vernon, R., 'International Investment and International Trade in the Product Cycle,' *Quarterly Journal of Economics*, 80(1966): 190–207.

Weber, Alfred, *Theory of the Location of Industries*, translated by C. Friedrich, Chicago: University of Chicago Press, 1929.

Young, Allyn, 'Increasing Returns and Economic Progress,' *Economic Journal*, 38(1928): 527–542.

Printed and bound by CPI Group (UK) Ltd, Croydon, CR0 4YY

22/10/2024

01777625-0015